THE CHAIN STRAIGHTENERS

THE CHAIN STRAIGHTENERS

Fruitful Innovation : the Discovery of Linear and Stereoregular Synthetic Polymers

Frank M. McMillan

M

First published 1979 by
THE MACMILLAN PRESS LTD
London and Basingstoke
Associated companies in Delhi Dublin
Hong Kong Johannesburg Lagos Melbourne
New York Singapore and Tokyo

Typeset by
Reproduction Drawings Ltd, Sutton, Surrey

Printed in Great Britain by
Unwin Brothers Limited,
The Gresham Press, Old Woking, Surrey

British Library Cataloguing in Publication Data

McMillan, Frank M.
The chain straighteners.
1. Polymers and polymerization–History
2. Synthetic products–History
I. Title
547'.8427'0904 TP1087

ISBN 0–333–25929–7

To the late
Sir Robert Robinson
without whose aid and encouragement
this book would not have been written

Contents

Foreword

The People of the Book—a Necessary *Élite*

Everyone knows how much Western industrial society has changed in the last century and by how much, for most people, the mechanics of living have been improved. Many also understand, in principle, that these improvements have been provided by the discoveries of science and by the work of industrial technologists turning new knowledge into useful productive techniques. To most people, however, this appears to result from a sort of self-perpetuating magic whose everlasting beneficity may be assumed. It is a particular merit of this book that it illuminates the process of discovery and its industrial development, particularly at the highest levels of decision making. At the same time it dramatises the actions of the chemical industry, in a period of rapid expansion, as it responds to a discovery whose impact was quickly perceived. At the highest levels we observe the action of those in decisive positions and the manner in which they determine the success or failure of major developments, and in the shadowy background we are dimly aware of the hundreds of research technologists who have the skill and dedication to make such decisions meaningful. To most of us, however, all this occurs in the dark distance from which no light emerges to compete with the entertainment-oriented information with which we are daily drenched. We may, therefore, reasonably ask who are these remote magicians, where do they come from and how do they set about their work? The book describes their actions and it is an object of this foreword to say something about the framework of their lives.

The *dramatis personnae* of Dr McMillan's book belong to a scientific and technological *élite*. By this we mean that they are people who, by a combination of suitable genetic endowment and a willingness to employ their talents for the mastery of complex and demanding problems, contribute more, and occasionally far more, to the material resources of society than most of their contemporaries. Such a distinction intentionally brackets together all the pure and applied scientists, technologists and engineers who together create useful knowledge and put it to work. Within each category there are outstanding individuals whose productivity and impact far exceed that of their colleagues so that great inequality is also the rule among the *élite* themselves. Readers of this book will be left in no doubt of the outstanding record of Ziegler and Natta, in providing the impulse for the great scientific and industrial developments with which the book is mainly concerned, and which was recognised by their Nobel award.

Once a useful discovery has been made industrial development takes over the demanding and expensive job of turning ideas and chemical reactions into useful products. In the early stages of this process publicity is often avoided, and when science history is written it is often so long after the events that it no longer reflects the original organisational framework in which it took place. It is a particular quality of Dr McMillan's work that it tells us about what happened next on a modern, international scale after the initial discoveries became known. On this count alone the publication of this book must be warmly welcomed. It opens a door giving a glimpse of the way in which the technological armies of the great multi-national corporations actually manoeuvre and work. Of this more later.

It cannot be doubted that a major part of the material improvements which have so greatly influenced the lives of ordinary people in the industrially developed countries have been generated by scientific progress and its industrial application. Few cases are known where societies have rejected the advantages attainable by technical progress when they had the capability of attaining them, and, although in 'advanced' countries a fair amount of noise is, and has been, made by those who affect to favour the 'simple life', few have voted for this option 'with their feet'.

The recent successes of applied science have, however, tended to obscure the rather specific conditions under which it can succeed. A technologically advanced society is an artefact of man's work, existing each moment in fleeting equilibrium with the forces of nature. It can never stand still. Sources of sustenance and power are discovered, harnessed, multiplied or exhausted in accordance with their inherent properties and the extent to which we are able to work in harmony with them. There is no way in which, over a significant period of history (say 50–100 years), an industrial society or its economic base can remain unchanged. A fossilised economy can only exchange goods and services with the active world on the basis of inferiority and supplication. So unfortunately it often turns out that the natural span of many jobs is less than a man's working life. The natural desire of many to keep them unchanged can only be achieved by allowing socially unnecessary work to be supported by the effective work force. No one is more affected by this turmoil than the technologists themselves who must often change their activities when existing projects have been taken to the limit of their utility. Thus because the option of a static industrial system is unattainable, such a society must either adapt itself continuously or decline. In the long run, without continuing innovation to meet new circumstances an economy must contract, not just relatively in a table of continuing expansion as with Britain today, but absolutely with a reduction in the sophistication of productive processes and an increase in actual poverty.

The process of adaptation is carried out by people, and, in the chemical industry, with which we are particularly concerned, by chemists, physicists and engineers. In this matter, however, the chemical industry does not differ fundamentally from other technological industries, it is just one of them—one of the most important. Without scientists they could not have started, and without great efforts from those able to come to grips with their technical problems they cannot continue. They inevitably consume a hierarchy of talents ranging, if you like, from Ziegler downwards, and, particularly when, as in this book, a field of great promise has been publicly demonstrated, competition between individuals and groups seeking to make a technical contribution is intense and national frontiers are no barrier. The results of these efforts provide patents, technological know-how and products. Of these the former perform the important role of allowing the publication of research while securing a dominating position for the enterprises which are first in the field. In this jungle of survival the common plodder can only perform a function like unto him who sells the tickets at the entrance to the olympic arena; and in the industrial laboratories, as in the stadium, great prizes are obtained mainly by those at the top. Lesser achievements are, of course, also viable, for it would be misleading to suggest that most industrial research is concerned with other than *relatively* mundane activities in support of existing processes and products, and the maintenance of efficient operations. But for this high standards are also demanded. In the last resort, if the *élite* of the laboratory and their colleagues in management and marketing fall below a certain level of performance the laboratory may become useless and the plant an object for industrial archaeology. Our industrial scene exemplifies the failures as well as the successes of technology and management. The patent which is too late is of little value and there is a time after which the development of a new process will itself be out of date.

Thus in scientific development work time may often be more important than money, and there is seldom a way in which the lack of talent and dedication in a research team can be compensated by greater numbers and lower wages. In Britain (and to some extent in other industrial countries) we need to understand that for applied scientists the notion of working a few hours each week on some sophisticated side-line hardly ever exists. We are not some sort of superior race, gifted to achieve in a few hours what lesser people can only do by long and concentrated effort. There are people on the other side of the world who will not agree to continue in a lowly role to which some of us would, unofficially, like to assign them. If we do not perform our share of the task of modifying our industrial society to changing situations then others will do it for us and in doing it they will change the balance of forces in the world.

For technological success generates a kind of power, which cannot be denied. Through its action the influence of national groups on the world stage imperceptibly shifts. If nations like Britain should wish to opt out of the competition, they will find that their future is increasingly determined by others. One part of this process is seen and generally ascribed to the arbitrary operation of the 'multinationals' who build up vital aggregates of technical know-how for particular industries. Through their research groups they create the technical capability to take advantage of scientific discoveries from any source. To the outsider, and even to many of those working in them, these great organisations seem remote and hostile, but their very existence is tied up with the sometimes embarrassing advantages of international rationalisation, and even more with the accumulation of relevant technical and scientific knowledge cultivated in large centres of research and development. They are therefore, for good or evil, large employers of our *élite*. Wherever, as in the drug industry, legislation and the readiness of the media to stress all failures, incompetently careless of the balance of truth, the cost of developing a new product escalates, thus intensifying the tendency of research to be concentrated in large groups. Only a few can then afford it. It is another important merit of this book that it throws further light on the way multinationals handle scientific information given in confidence. Occasionally standards fall below a desirable level but at other times they are high. Indeed, we see that in two cases particular corporations adopt a high ethical level in handling such information, in one case through a perceived moral code, and in the other by close adherence to an unenforceable verbal agreement. Such actions which in my experience of work in the Shell Company are not untypical, offer the hope that it may be possible to establish some more widely agreed international codes of behaviour for their operation. Indeed their present actions, often seemingly irresponsible, are generally less arbitrary than many interested parties would like to think. An unsuccessful laboratory or a loss-making factory is what it is, and likely to be closed whoever owns it, and the factors which lead to failure will often operate independently of ownership. However the multinationals are often very coy about providing the experience and figures on which their decisions are based. In some cases widespread social trends (as in the demand for ever-shorter hours of work) may undermine success. In other cases, especially in the advanced technologies, only a very small proportion of the population is responsible for maintaining standards of operation and developing the products and processes for the future and where these fail many others become involved.

There are some signs that the size of the pool of available talent and motivation for this *élite* may be falling in the U.K. Not, perhaps, at the top levels of academic science where freedom to select one's own problems and open competition for public and professional recognition are still a strong spur, but at lower levels, in universities, and in the economically more sensitive areas of industrial science. Here there are two aspects. In the first place many industrial countries in recent years have been affected by a drop in relative and absolute numbers entering university to study science, technology and engineering courses, though happily there is recent evidence that the trend is not continuing—but there is still some way to go. In the U.K. the proportion of applicants seeking university places in physics and chemistry has fallen from 6% to 2% between 1963 and 1978 and though this is a fraction of a rapidly rising total, in the case of chemistry there was also an absolute decline in numbers (*Nature*, Oct. 19th, 1978).

Thus although it is argued by some educationists that the general standards reached in the schools (as represented in the U.K. by total 'A' levels) is increasing, it appears that certain demanding scientific subjects are being avoided and the present figures are, therefore, not comparable with previous ones. Probably more important, though also more arguable, are concepts which aim at levelling the performance of children in schools, so that all may be as equal as possible not just in opportunity but in attainment. According to this philosophy it is even desirable to hold back the more talented in the hope of achieving a slight improvement in the weaker performers. Naturally, this trend is often opposed by the parents of talented children who may or may not be in a position to influence events. The same principles are, further, not generally considered applicable to the arts, e.g. music and ballet for which special schools exist to maximise the potential of promising boys and girls. Nobody wants to go and see the common man performing on the concert platform. In the more general application, however, the equalitarian philosophy is generally bolstered by the rather optimistic assumption that society can indefinitely provide a good life for all on the basis of very moderate effort and much spare time.

Such a philosophy if followed through will hardly provide us with the necessary technological *élite*. To this end some young people must be trained and motivated to efforts which much exceed the norm and compete with those available elsewhere. No doubt there will always be a few who will respond to the challenge of difficult work and to the satisfaction of finding a viable path through a maze of the difficulties, but what happens if at the end of the day the job is still unfinished and the wife rings up and says there's a leak in the watermain and the plumber hasn't come (or cannot be afforded)? It would seem unwise to rely entirely on willing horses to rescue us in so vital a matter.

More generally people will respond to the incentive of money and fame. For the industrial technologists today, however, unlike their Victorian predecessors, a fortune is a most unlikely outcome, and the very nature and complexity of their work makes them unattractive to the media. Even their best friends (outside their profession) will have little idea of what they are doing and the impact of their work on society may emerge from a vast organisation in a wholly impersonal form. Indeed the comprehension of their work by the community is only too likely to be further limited by the (almost inevitable) under-representation of industrial technologists in public life. For most of them the main driving force is limited to that of recognition and promotion by their employer.

Let us consider the case of the discovery of Polythene by I.C.I. at Northwich, the history of which is now fairly well known and shortly summarised by Dr McMillan. This was an outstanding achievement both of scientific innovation and of recognition. The problem of recognising the utility of a novel product always seems easy with hindsight, but consider the position of scientists gazing at a bit of (probably) discoloured wax in the early 1930s, when nothing like Polythene was even known to exist! Indeed it will be clear from other parts of this book that even *10 years later,* when the high-pressure process had been established, other companies had difficulty in appreciating the possible importance of similar products. However the few men concerned in I.C.I. were equal to the occasion, and achieved what is (arguably) the commercially most important example of scientific innovation in Britain in the last 50 years. The result contributed significantly to Britain's military technology in World War II and to its balance of payments during the lifetime of the patents. We were also able to export equipment and know-how for Polythene plants all over the world. Naturally other countries and companies now compete in the field (as we compete in products invented elsewhere) but the upshot is a world-wide production of high-pressure Polythene in 1978 of some 12,000,000 tons and an annual turnover above £3,000,000,000. This exceeds that of high-density polyethylene by the Ziegler and Phillips processes and polypropylene put together. However the men who first made it possible have remained without public recognition outside the industry* either from the scientific or the civil establishments, although our honours system provides encouragement for many (in our view) less vital activities. Of course these same men generally had rewarding careers in their company and several rose to high office.

However, although we may properly reproach our rulers for prejudice and technological delinquency, it would be fair to admit that certain problems of recognition do arise. In the first place with Polythene it took 10–20 years before it was clear that a product of such major and permanent international importance had been produced, and by this time hundreds and perhaps thousands of scientists and engineers had made their varied and sometimes critical contribution to the development of commercial processes. Indeed this is typical of discoveries in the chemical industry whose importance is seldom realised at first, though it was Ziegler's and Natta's good fortune to be the exception to this rule. Further, industrial research is mostly

*Dr J. Swallow received the Swinburne award of the Plastics Institute.

carried out, especially today, in directed teams so that in the end it may not be so easy to determine true inventorship, especially when the actual originator of the new chemistry was not the one who first recognised its importance. After some time has passed, when patents have been published and commercial secrecy relaxed, it is generally found that more people thought of the good ideas than the bad ones.

However, when all is said and done, these difficulties are not insuperable and where there is great merit a greater effort should have been made to recognise it. In the meanwhile we need to appreciate that the leading industrial innovators, unlike their opposite numbers in entertainment and sport, can expect neither publicity nor the opportunity to generate a large income in countries with a favourable level of taxation. To be fair here one should say that the position in America is often somewhat better than in Britain, though I do sometimes wonder whether, in equity, Ziegler does tower above Hogan and the Phillips group to quite the extent that corresponds to the consensus of scientific opinion. In any case for us in Britain the education and encouragement of industrial scientists should be the subject of particular concern.

As already noted, a feature of a great technological innovation is that, whether or not its importance is recognised quickly, it provides new and useful developments continuing far into the future. This applies to Ziegler perhaps even more than to I.C.I., because his discovery was more general. As the reader will discover from the main narrative, in the period following his invention, with one or two exceptions, of which Hoechst was perhaps the most notable, Ziegler's polyethylene had a hard time in competing with those from the Phillips process, which at present have a greater installed capacity for the high-density type of polyethylene. However, this situation appears to be changing. Ziegler's chemistry was complex and the number of significant variations coming within his possible catalyst systems was almost infinite. This made the selection of the initial systems for commercial development more difficult, though the ultimate possibilities after extensive research are much greater. These longer range possibilities are still being developed. In the polymerisation of ethylene Solvay pioneered a 'high-mileage'-supported Ziegler catalyst which provided a much higher yield of polymer per gram of catalyst and so reduced the problems of polymer contamination and eliminated the requirement for an expensive purification process. Other versions of such high-activity catalysts, mostly falling within the definition of Ziegler's original patents, have been described in the recent scientific literature. So now Ziegler-type processes may actually become cheaper than the high-pressure system, as the early licencees had hoped. As a result we now find Union Carbide announcing a major 100,000 ton plant for making low-density polyethylene by a low-pressure process and in doing so they join du Pont of Canada who have been making a somewhat similar material for more than 15 years. However, as is always the case with polymers, the low-density polyethylenes made by the I.C.I.-type and Ziegler-type processes are not the same. In the high-pressure process the polyethylene, whose crystallinity is controlled by short-chain branching also has long-chain branches. In the free-radical mechanism this cannot be avoided, but in the low-pressure (Ziegler, Phillips and other) processes it is possible to produce materials of the same controlled crystallinity but without long-chain branches, thus giving a new twist to McMillan's chain-straightening concept! Since the new products have different properties, improved in some respects if not in others, they are expected to succeed in certain markets. So there is currently much fluttering in polyethylene dovecotes as Ziegler's revolution marches on.

It is a great merit of Dr McMillan's book that it displays the processes by which scientific discovery and technical innovation interact in a modern environment. In this foreword I have tried to describe the position of industrial scientists against a wider canvass, especially from the British viewpoint (for which I hope Dr McMillan will forgive me). I also underline the consequences of neglecting the vital functions which they undertake.

Perhaps I might appropriately close on a personal note. During the period when Dr McMillan describes events at Petrochemicals Ltd, I was also on site. However at the time the company as a whole was 'in the red', so that it did seem very important that the one part of it (the polystyrene plant) which was making a profit, should stay that way. This kept me and my few colleagues (including the Erinoid group then at Stroud) more than busy. However, on the day when Bernard Wright fed propylene to his reactor he rang me up, as well as Dr Borrows, and I came across to have a look at his lumps of polymer. Little did I realise that I was gazing at the biggest bit of polypropylene in the world!

January 20th, 1979

R. N. Haward
Department of Chemistry
University of Birmingham
P.O. Box 363
Birmingham, B15 2TT

Introduction and Acknowledgements

> There is more than one side to the role of men in science, who are not only carriers and shapers of ideas but also personalities who, for better or for worse, have often moulded the course of events more through their strength and idiosyncracies than through their science.
>
> —Tjeerd H. van Andel

Science and technology usually advance by small, almost imperceptible, steps. Only after the slow accumulation of many such steps do we find ourselves on wholly new, and notably higher, ground. But once in a great while a single discovery is so significant that at one bound it lifts an entire field to new levels. Occasionally the key discovery is the brainchild of a single individual. If that person is unusually fortunate, his achievement is recognised during his lifetime and is rewarded with various honours, perhaps even with the Nobel Prize. In very rare instances, it even makes him rich.

This is a story in which all those things came true. It is the story of a key discovery in polymer science: a discovery that, in turn, triggered a rapid-fire sequence of related advances by people in several different countries, working partly in concert and partly independently or in competition. This floodtide of discovery has left its record in thousands of pages of scientific papers and in thousands of patents. But, as always, there is a personal story behind the scientific record. The principals, Karl Ziegler and Giulio Natta, are now famous personages and Nobel laureates, but that does not make them less human. They also had numerous predecessors and contemporaries, equally human, whose contributions have not always received as much recognition as they and some others have felt they deserved.

The present narrative is not intended to extend the already extensive scientific record, nor to redress any imbalance of credits, but to tell the human story—to describe the people who made the discoveries, their backgrounds, where and how they lived and worked, and their interactions with one another. Out of all this should come some insight as to how discoveries such as theirs come to be made, and some appreciation of the long-range consequences of what they did.

But there is an additional objective: one having to do with innovation and creativity. The quest for the well-springs of 'creativity' of one sort or another has occupied many writers; my concern is more specifically with what I shall call *'fruitful innovation'*, meaning a basic discovery that is rapidly expanded by corollary findings and technological developments to the early benefit both of man's knowledge and of his well-being.

The story at hand provides a textbook example, or, rather, several examples. Many scientists have perhaps been fully as creative as those named here, but few have seen their findings expanded and developed so fruitfully and so quickly. The frequency of such an outcome might be improved if case histories were examined to reveal key factors responsible for such early fruition.

To what extent the critical factors in this instance might apply in others, and which might be amenable to control, can of course be judged only after they are identified. The author's conclusions on these questions are drawn in the hope of being, if not definitive, at least provocative.

Sources

The primary source of material for this book was a series of personal interviews with people who were either principals, participants or first-hand observers of the events described. Others, unavailable for oral interviews, were consulted by correspondence. These direct sources were supplemented by reference to the literature, to contemporary press accounts, and to special documents such as doctoral theses and transcripts of legal depositions. The author's own recollections were relied on sparingly in regard to certain minor incidents and details.

Those familiar with the subject may find the low frequency of patent references surprising, since the commercially significant discoveries were usually published first and, in some cases, solely in the form of patent specifications. Patents pertinent to stereoregular polymers number in the thousands, and their pages in the tens of thousands. Nevertheless, I have made very little reference to patent disclosures in this book, nor have I discussed patent priorities and conflicts, save where the disclosures are the only source and the conflicts an integral part of the story.

This restraint stems from several reasons: (1) patent disclosures generally present a few islands of fact in a sea of exaggeration, and are therefore at best a turgid, and at worst a misleading, guide to what actually happened; (2) nearly all of the information needed for our purpose is available from less ambiguous and less biased sources; (3) excellent summaries and reviews of the patent literature have been published*; and (4) the critical evaluation of patents and the issues of validity, priority and infringement are touchy professional questions best left to professionals in that field, especially since some major issues are still unresolved.

No account of human activities, based in part on human memories and personal viewpoints, can be free from error or bias. The author's apologies are offered for those omissions and errors that could readily have been avoided and for any of his own prejudices that show through his veil of impartial reporting. The situation faced by author and reader alike has been put forcibly by Jewkes *et al.*†:

> So, too, the writings on invention are of extraordinarily mixed quality. There seems to be no subject in which tradition and uncritical stories, casual rumours, sweeping generalisations, myths and conflicting records more widely abound, in which every man seems to be interested and in which, perhaps because miracles seem to be the natural order, scepticism is at a discount.

**See*, for example, C. E. Schildknecht, *Allyl Compounds and their Polymers (including Polyolefins)*, Wiley-Interscience, New York, (1973); Kennedy and Tornqvist (Eds), *Polymer Chemistry of Synthetic Elastomers*, Wiley-Interscience, New York, (1968).

†Jewkes, J., *et al.*, *The Sources of Invention*, Macmillan, London (1961).

Acknowledgements

Any list of obligations is bound to be incomplete, but I wish to express my particular thanks to those named below for the assistance and courtesies that made this book possible.

(i) For personal interviews and first-hand descriptions that contributed vitally to the authenticity and human interest of the narrative:

Mr John Aldersley, Mr Leonard Batten, Dr E. T. Borrows, Mr Leif Christiansen, Mr Aldo DeBenedictis, Dr John DeNie, Dr E. W. Duck, Dr Lawrence Forman, Dr Fred Foster, Prof. Paul Flory, Dr Arthur Glasebrook, Dr Morton Golub, Dr Michel Goppel, Dr Leslie Holliday, Dr Sam Horne, Dr Maurice Huggins, Mr Paul Johnstone, Mr Adrian Koeleman, Dr Art W. Langer, Dr R. N. Legge, Dr E. Lelyveld, Dr Heinz Martin, Prof. C. S. Marvel, Prof. Maurice Morton, Prof. I. Pasquon, Prof. Piero Pino, Dr G. E. P. Smith, Jr, Dr Fred Stavely, Dr Erik Tornqvist and Dr E. J. Vandenberg.

(ii) For special assistance and encouragement:

Prof. R. N. Haward, Prof. Herman Mark, Sir Robert Robinson and Prof. Calvin Schildknecht.

(iii) For information provided through correspondence and publications:

Dr Case Baas, Mr John Bell, Dr James D'Ianni, Mr J. P. Hogan, Prof. Giulio Natta, Miss Leora Straka and Prof. Karl Ziegler.

(iv) For helpful criticism and manuscript correction:

Dr Fred Condo, Mr Fred Hilmer, Mr Robert Martin and Dr Paul Williams.

(v) For patience and encouragement:

(Mrs) Frances J. McMillan.

1 The Stage is Set

In 1953, in Germany, Prof. Karl Ziegler discovered a process for making a new type of high polymer that was recognised immediately as a material of great scientific interest and practical utility: linear, crystalline polyethylene. In the following year he made a solid polymer of propylene; however, he was anticipated in this discovery by Prof. Giulio Natta in Italy who independently discovered this remarkable new polymer: crystalline polypropylene. Natta conclusively proved the unprecedented regularity of its molecules and helical form of its crystals. In 1963, these two distinguished scientists shared the Nobel Prize in Chemistry for having jointly created the science of 'regular' polymers and stereospecific polymerisation.

The opening of this rich new field of science and technology generated feverish research activity and plant construction all over the world. Science has been significantly enriched by the new chemistry thus developed, and most of us benefit daily from the many useful materials made with the aid of that chemistry.

It seems important to ask, regarding such phenomenally fruitful innovations, what characteristics and conditions were responsible for them—both for the initial success and for the almost explosively rapid developments that followed. If the key factors can be identified, knowing what to look for should help the cause of those few who can aspire to similar accomplishments and the understanding of the many who can benefit from them.

In telling the story that is the subject of this book, we shall look for the decisive events and seminal influences in the lives and labours of the major innovators. But these inventors did not work alone or live in a vacuum. We shall therefore have something to say about their predecessors and their contemporaries—collaborators, supporters and critics—and about the nature of the times in which they worked. All had their influences, and all contributed to the progress of events that culminated in the discovery of the linear and stereoregular polymers.

Our story is primarily about people rather than their scientific work *per se*, the latter having been amply documented in the scientific literature*. But the people can be understood fully only with reference to their work, and the latter must be described in terms of the basic principles of polymer chemistry. A brief summary of the relevant highlights of that fascinating field may therefore be useful as a refresher or an introduction to some readers, even though it may draw smiles from the expert and groans from the novice. Both may skip lightly through the indented passage that follows (and others that appear later)–the one because it is familiar ground, and the other because it may exceed his interest and is not indispensible to following the narrative.

But for those who have had a modest exposure to physical science–enough to be intrigued, or at least not frightened, by it–renewing or extending a nodding acquaintance with polymers may keep them from nodding over the rest of this story. It can also add to an appreciation of the wonderful world of nature.

For polymer science is not a remote or esoteric subject. Both ancient and modern man has lived in a world in which the material things seen and touched and used are mostly high polymers, i.e. they are sub-

---

*From a literature of tens of thousands of pages and thousands of patents in many lands and languages, we may note the following as providing good introductions and surveys:

Schildknecht, C. E., 'Stereoregulation and Stereoregular Polymers', *Polym. Engng & Sci.*, 6 No. 3, 1 (1966)

Schildknecht, C. E., *Allyl Compounds and Their Polymers (Including Polyolefins)*, Wiley-Interscience, New York, Ch. 2 (1973)

Ziegler, K., *et al.*, 'The Mülheim Normal-Pressure Polyethylene Process' (in German), *Angew. Chem.*, **67** No. 19/20, 541–636 (1955)

Ziegler, K., 'A Forty Years' Stroll Through the Realms of Organometallic Chemistry', *Adv. Organometal. Chem.*, **6**, Academic Press, New York (1968)

Natta, G., 'Macromolecular Chemistry', *Science,* **147**, 261–72 (1965)

Gaylord, N. and Mark, H., *Linear and Stereoregular Addition Polymers,* Interscience, New York (1959)

Kennedy, J. and Tornqvist, E., *Polymer Chemistry of Synthetic Elastomers,* Wiley-Interscience, New York (1968)

Mark, H., 'Giant Molecules', *Time* (1966)

Raff, R. and Doak, K. (Eds), *Crystalline Olefin Polymers,* Interscience, New York (1964–5)

stances composed of enormously long, threadlike molecules. Early man used the natural polymers he found all about him: tree trunks and sticks; grasses and reeds; hides, bones, sinews and hair; fibres and feathers. Modern man sees and uses all these plus a vast and varied array of other materials, of which polymers still comprise the overwhelming majority.

Many of the polymers we use are now synthetic: fibres, films, plastics, coatings, rubber, etc. Already we use a greater volume of plastics than of iron and steel. In fact, as Prof. Giulio Natta, one of the principals in this story, has said: 'If our age were to be named for the materials that characterise it–this might be known as the age of plastics. For plastics, made of synthetic giant molecules, have become a dominating influence on modern industrial society.'

Still more importantly, high polymers are the very stuff of life. We ourselves are composed mostly of high polymers of extraordinary complexity and exquisite specificity: muscles and nerves and tendons; enzymes, skin, hair and nails; even our brains and the substance of our memories. So we have lived intimately with polymers from the beginning of life, and understanding them is vital to understanding ourselves. But real understanding has come slowly, because polymer science is relatively new. The empirical arts of use preceded scientific understanding by millenia in the case of natural polymers and by decades even in the case of synthetics.

What is a 'high polymer'?

The word 'polymer' is defined by its derivation–from the Greek words *poly*, meaning 'many', and *meros*, meaning 'parts'–a large molecule built up by combining many repetitive units (monomers: 'single parts'). A 'high' polymer is simply one combining a high number of units to a high molecular weight.

The manner in which the monomer units are combined is a question that baffled chemists for generations. They were led unwittingly down false trails by the 'colloid chemists', who had successfully explained emulsions and suspensions in terms of aggregates of smaller particles, and who proceeded naturally to extend that concept to natural high polymers–rubber, proteins, cellulose, etc. As Olby has put it: 'By common consent the path to the study of . . . the structure of natural products lay through the pastures of the colloid chemist.'

Unfortunately, the 'aggregate' approach proved in this case to be, not the high road to understanding, but a sterile bypath that delayed the arrival at true insight for at least a decade, perhaps more.

Birth of polymer science

In the early 1920s, the Swedish scientist, The Svedberg, ended this 'mediaeval' era of polymers and began the enlightenment. With the aid of his powerful new instrument, the ultracentrifuge, Svedberg showed that polymers were not simply irregular physical aggregates, but true molecules of undreamed-of size. He measured molecular weights of as much as one million, in contrast to the 100, or less, characteristic of most ordinary compounds. This meant that a single molecule could contain, not just a few atoms, but hundreds of thousands.

After this innovative concept was finally accepted (with great reluctance and resistance, even though Svedberg was awarded the Nobel Prize for this work in 1926), the question remained as to how such vast numbers of atoms were combined. The answer was provided by a man who had welcomed Svedberg's measurements because he already believed in giants, at least where polymers were concerned. That man was Prof. Hermann Staudinger, the outstanding German organic chemist whose many pioneering contributions make him the 'grandfather' of modern polymer science. He argued that, in high polymers, the atoms are linked, not in blocks, rings or networks, but in enormously long chains.

This fascinating proposition got an even worse reception than Svedberg's. It is hard for us today to imagine waxing emotional over the size of molecules or their shape, but at a lecture in Zurich, Staudinger met with such vociferous dissent that the meeting ended in uproar, with Staudinger stoutly holding his ground and saying, with Luther: 'Here I stand, I cannot do otherwise!' Resistance was so severe and prolonged that Staudinger had to wait 33 years, until 1953, for his Nobel Prize (perhaps he simply outlived some of the diehards on the Committee).

It was in that same year, 1953, that Ziegler discovered linear polyethylene, Hillary and Tenzing climbed Everest, and Watson and Crick discovered the double helical structure of the DNA molecule. The 'instant hero' status won in these latter-day triumphs must have brought an ironical smile to the old pioneer who could justly claim that his work had made at least two of them possible.

Carbon chains

Most polymers are organic, i.e. compounds of carbon, because carbon atoms have an outstanding proclivity for linking with each other in chains:

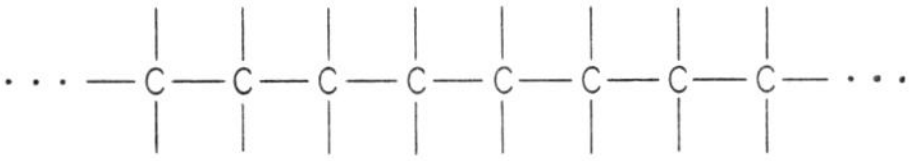

In many polymers, both natural and synthetic (e.g. cellulose, nylon), and in 'living' polymers such as proteins, other atoms (oxygen, nitrogen, sulphur) are interspersed with carbon, and various other groups may be attached at the side of the main chain. In fact, Staudinger added to his long-chain theory the controversial proposition that the only difference between 'living' and 'non-living' molecules is their size. The heretical implication that life might then be created spontaneously (or deliberately!) roused a theological storm which modern biological science has not yet completely stilled.

Our case is simpler, because nearly all of the polymers mentioned in these pages are hydrocarbons. That is, they consist of carbon chains with only hydrogen or simple carbon-hydrogen groups attached to the side of the chain, for example:

$$\cdots -CH_2-CH_2-CH_2-CH_2-CH_2-CH_2-CH_2-CH_2-CH_2-CH_2-CH_2- \cdots$$

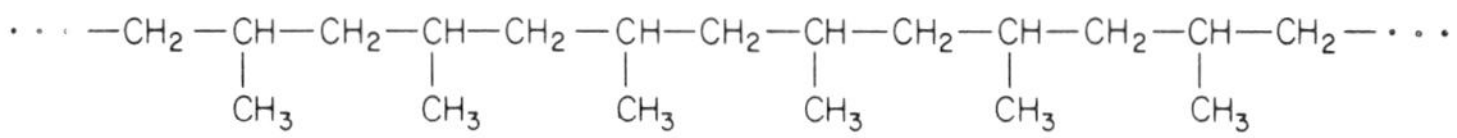

Even more simply, when the chain is built up by repeated additions of a single hydrocarbon molecule, the chain may be represented schematically to reflect that fact:

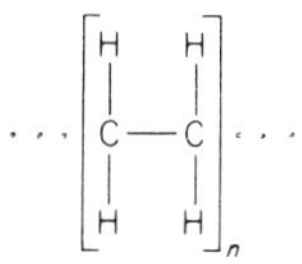

polyethylene

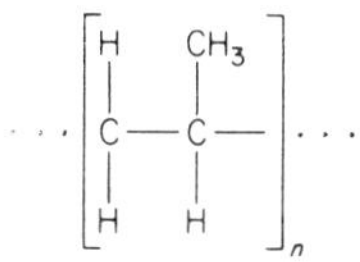

polypropylene

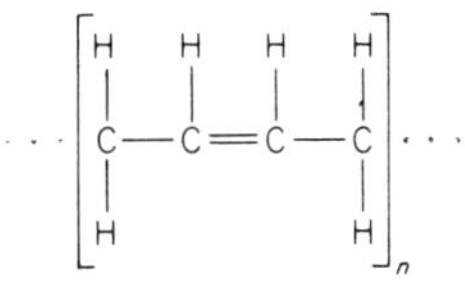

polybutadiene

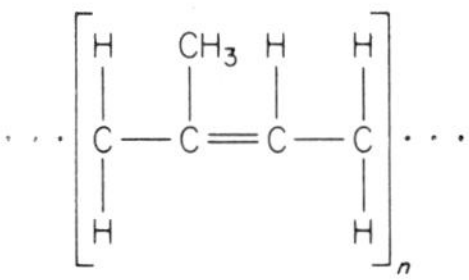

polyisoprene

In these schematic formulae, n, the number of repeating units, is a large number, typically from 1000 to 10 000. The chains are shown for simplicity in a 'straightened out', or extended, position, but their natural tendency is to kink and curl, so that a high polymer is actually a dense tangle of twisted and intertwined chains:

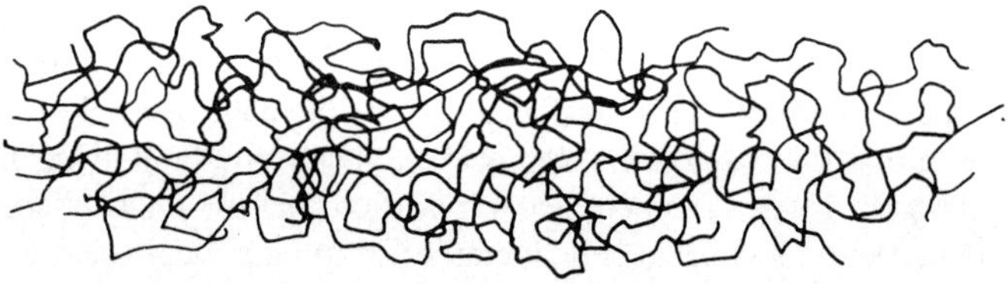

An additional complexity arises from the fact that the molecules may not be single straight chains, but ones with branches, either short or long:

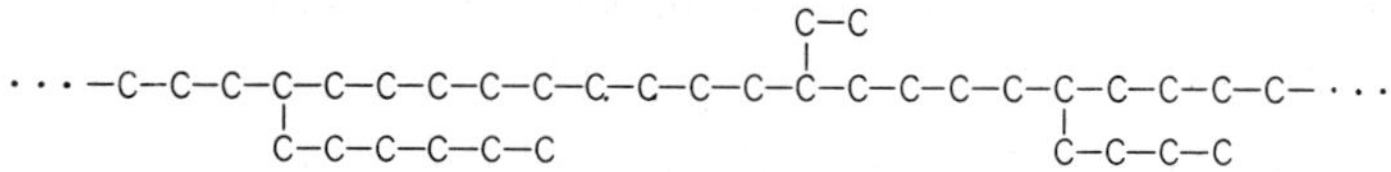

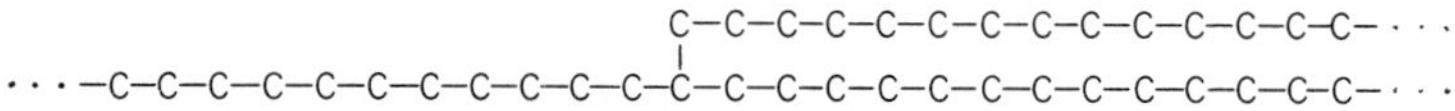

All of these differences affect the behaviour of the molecules and the bulk properties of the high polymer.

The achievement and the fascination of polymer science is its ability to explain the known physical properties of high polymers in terms of their molecular structure, to predict previously unsuspected properties and, finally, to synthesise new polymers with particular desired properties ('tailor-made molecules').

It is noteworthy that at every step in the progressive elucidation of the structure of natural high polymers, an additional unsuspected degree of order and regularity was revealed. First, they were found to be true molecules, not loose agglomerates. Then, the molecules were shown to be, not random linkages, but long, threadlike chains. And, finally, as we shall see later, it was discovered that the chains are constructed very selectively so as to have highly specific steric ('stereoregular') configurations.

Man's first attempts to duplicate nature's high polymers, beginning long before he knew their molecular structure, fell far short of his goals. Although useful products were made and marketed under such optimistic labels as 'artificial silk', 'synthetic rubber', etc., they were, in fact, markedly different from and, in many respects, inferior to, the natural products they sought to emulate. This situation persisted for many years, despite continued efforts and despite the fact that the proper molecular weights (chain lengths) were achieved.

It was only a few decades back that it was proven that these differences were due to the exquisite regularity of the high polymer molecules fashioned by nature and that the lack of such regularity was responsible for most of the practical shortcomings of the synthetic analogues (just as a felt, which is a mat of randomly arranged fibres, is a useful cloth but does not duplicate the properties of a regularly woven fabric). And it was only at the time of our story that it was discovered how to achieve the requisite high degree of regularity in synthetic structures and how to use it both to duplicate a natural high polymer and to create new, crystalline, high polymers that nature had never made.

The importance of having highly regular, unbranched chain structure lies primarily in the fact that such molecules will fit together to form tiny bundles, or 'sheaves', that are actually microcrystals, with a very high degree of local order. By tying up otherwise flexible chain segments in rigid structures, these microcrystals ('crystallites') markedly increase the hardness, heat resistance (softening temperature) and solvent resistance of the polymer, and thus greatly increase its usefulness for many purposes. This may be represented in highly schematic fashion as follows:

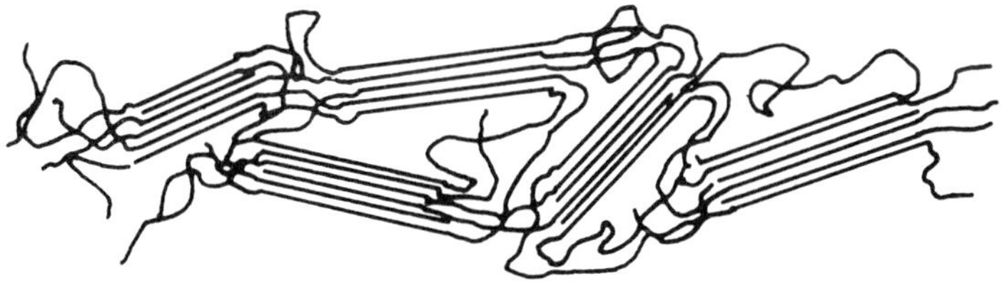

The 'state of the art'

At the time that our story really gets under way, the beautiful regularities of natural polymer molecules were only partially revealed, and their duplication in the laboratory seemed an impossible dream. True, Wallace Carothers, of duPont's pure-research laboratory ('Purity Hall' to the cynics), had confirmed Staudinger's long-chain theory in the only way convincing to an organic chemist—by synthesising long chains in a controlled, stepwise procedure. Nevertheless, that theory was still a relatively new doctrine, and backsliding was not unknown*.

By the early 1950s, the structure of natural rubber was reasonably well understood, but that of other important natural high polymers, such as proteins, nucleic acids and cellulose, constituted hot but baffling research problems for some of the best scientific brains in many countries.

The synthesis of most high polymers, at least on any large scale, was a semiempirical and often imperfectly controlled process. Each batch of a synthetic high polymer had to undergo elaborate performance tests, rather than the simple analytical checks that sufficed for ordinary chemicals. Production people chafed at this requirement, and those of us who were in research had to acknowledge that our reactions were not manageable enough to determine precisely either the length or the shape of the polymer chain, although both were undoubtedly important. Considering the number of possible variations in structure, it seemed doubtful that we had ever made two molecules that were strictly identical.

Impressed with these complexities, I once made the prediction that we might aspire eventually to gain control over chain length, but could not realistically hope ever to match Nature's beautiful control over chain shape. Fortunately, that prognostication was little noted nor long remembered, and the outcome was exactly the opposite.

We may well ask, now that the scientists we may irreverently dub the 'Chain Straighteners' have actually achieved near-perfect shape control for many polymers, whether the seemingly simpler problem of length control might not be solved also. A reasonable answer would be: 'Yes, when the incentive becomes great enough.'

---

*During World War II, when some of the best research and engineering talent in the U.S. was striving to produce unprecedented quantities of a new high polymer that would substitute for (although not duplicate) natural rubber, I was dismayed to hear the coordinator of the huge government-sponsored research program say that he was not so sure that the 'micelle' (aggregate) theory of the colloid chemists was not right after all! Somehow, despite this appalling heresy, the plants got built and the war won.

2 The Shadows Before ('Prediscoveries')

History has no beginnings. –Will Durant.

Vikings crossed the Atlantic before Columbus, and Langley built an airplane before the Wright brothers. Darwin stood on the shoulders of Mendel and Malthus. It does not surprise us today, as soon as a scientist's work has been hailed as an original and major contribution, to begin hearing of others whose less-heralded labours paved the road the now-famous then pursued to glory. Such 'prediscoveries' shine more brightly in the afterglow of later knowledge, and the lens of hindsight may be employed to magnify their significance at the expense of the culminating discovery, even to the point where the latter may be made to appear less a great leap forward than merely a last logical, and even inevitable, step.

The discovery of stereoregular polymers is no exception. It was foreshadowed by certain events in polymer science and by aborted discoveries that, like minor tremors preceding a major earthquake, can be seen afterwards as signalling and triggering earth-shaking events. Karl Ziegler and Giulio Natta are justly famous for their epochal discoveries: Ziegler for his linear, crystalline polyethylene, Natta for his isotactic, crystalline polypropylene and other stereoregular polymers*. Yet Ziegler was not the first to make a linear polyethylene, and stereoregular polymers were postulated, prepared and published prior to Natta's work.

Linear, crystalline 'polyethylene' was, in fact, made half a century before Ziegler by another German, named von Pechmann. He made it

* See page 10

by decomposing diazomethane – an elegant but impractical process. It is elegant because it is capable of yielding *only* a completely linear polymer, since the methylene (CH_2) groups are added one by one to the growing chain:

$$\underset{\text{diazomethane}}{CH_2N_2} \longrightarrow \underset{\text{polymethylene}}{CH_2CH_2CH_2CH_2CH_2CH_2CH_2CH_2-\ldots} + \underset{\text{nitrogen}}{N_2}$$

Considering its origins, the product is properly called *polymethylene*, but it is structurally indistinguishable (ignoring end groups) from a perfectly linear polymer of ethylene.

Since this was purely a laboratory method, the product was regarded as merely a laboratory curiosity, and neither von Pechmann nor anyone else was stimulated to seek a practical route for duplicating this entirely novel product. The reasons for the lack of recognition and follow-up that characterised this and numerous other 'prediscoveries' will be discussed in our closing chapter, but in von Pechmann's case it was mostly a matter of simply being far ahead of his time.

High-pressure polyethylene (Polythene)

A different type of polyethylene, 'branched' polyethylene†, did become a commercial reality and an enormous success twenty years before Ziegler's, as a result of a remarkable combination of circumstances in the laboratories of Imperial Chemical Industries (ICI), England's largest

---

*Terminology in the field has proliferated but has not always been consistent. 'Official' definitions are now being established to make obsolete some of the early, more graphic expressions (e.g. 'stereopolymers', 'stereospecific polymers'). We shall employ the minimum necessary number of special terms, defined informally as follows:

'*Regular polymer*': one composed entirely of sequentially repeating units. This general class includes both linear polyethylene (i.e. poly $-CH_2-$, but not branched (high-pressure) polyethylene) and the stereoregular polymers.

'*Stereoregular polymer*': a regular polymer in which steric order (i.e. with respect to *D–L* or *cis–trans* isomerism) is observed. This includes both the 'tactic' polymers and the 'stereorubbers' that will be discussed later in more detail.

'*Stereospecific polymerisation*': a process which produces stereoregular polymers.

†*See* diagram on p. 6.

chemical company. The story of that discovery has been told well and often, although with interesting variations, by some of the principals and other chroniclers*. But it deserves brief recounting here because it had a great deal to do with setting the stage for the advent of the linear and stereoregular polymers. It is also valuable as an outstanding case history in our search for the origins of fruitful innovation. What J. C. Swallow, who shared in the discovery of high-pressure polyethylene, says of that event could also be said of many of the later discoveries:

'As time goes on the story of inventions and discoveries tends to become idealised and over-rationalised and presented in a way which suggests that everything follows logically from the first experiment. History shows that this rarely happens and that chance always plays a big part. What does, however, seem to be important is the significance of the recognition of a discovery which may often be more important than the discovery itself.'

More often than is usually recognised in the history of high polymers, the invention of specialised equipment has opened up and paved the way for major advances. Polymer science itself began with the building of Svedberg's ultracentrifuge and advanced with the development of x-ray and infrared spectrometers. High-pressure polyethylene was made possible by the development, by Bridgman in the U.S. and Michels in Holland, of equipment and techniques for working at previously unattainable pressures.

The pioneering work of Prof. P.W. Bridgman at Harvard on ultrahigh-pressure phenomena convinced J.C. Swallow and N. W. Perrin of ICI that a new frontier had been opened for chemical research. This view was supported by Sir Robert Robinson, a distinguished English organic chemist and (later) Nobel laureate, who was then a consultant for ICI. He visited Prof. Michels at Amsterdam University and became convinced that practical equipment could be designed for operation at pressures of 20 000 or even 40 000 pounds per square inch.

In 1932, 'a bleak time in this country', Swallow vividly recalls, he and Perrin wrote a memorandum that inaugurated a research program of searching for new phenomena in chemical reactions under extreme pressure. Fifty different suggested reactions were tested without success. But one of the failures resulted in a striking discovery through a most remarkable series of coincidences.

**See*, for example, Swallow, J. C., *Polythene*, Ed. Renfrew and Morgan, Interscience, New York, Ch. I.

Sir Robert had made one of the fifty suggestions: to try forcing benzaldehyde and ethylene to react. This experiment was made in March 1933, by R. O. Gibson. According to Swallow, the expected product was 'propiol lactone', but Sir Robert said later that he had expected paraethyl benzaldehyde. However, neither was formed, and the benzaldehyde was recovered unchanged. But on opening the pressure vessel, a thin coating of white waxy solid was found on the walls.

This solid obviously had to be some sort of a polymer of ethylene, so the experimenters allowed themselves to be distracted from their original goal and made a second test with ethylene alone. This time a violent decompositon occurred that caused 'bursts in the joints, tubes and gauges'. The alarmed experimenters concluded that prudence was the better part of scientific curiosity and accepted a delay while properly barricaded facilities were constructed.

Meanwhile, other reactions were studied, but without any exciting results. It was not until 1935 that work with ethylene was resumed, using new equipment designed with the help of Prof. Michels. This time the reactor pressure dropped steadily, indicating rapid reaction. Additional ethylene was pressed in to keep the reaction going, and when the vessel was opened, solid polymer was again found. But there was only a few grammes of it, a small fraction of the weight of ethylene that had disappeared. Leakage must have been responsible for most of the encouraging pressure drop! That leak was actually a very happy accident, although no-one supposed so at the time.

That first tiny sample was enough to convince the research people that they had 'an exciting new material', but research people are congenital optimists. More impressive is the fact that ICI management made a commitment to a major development program at that point—a time when there was only eight grammes of the product in existence!

The decision was obviously bold, but it soon began to look positively foolhardy. For the next several months, all attempts to duplicate the polymerisation reaction and make more material were uniformly unsuccessful. With the reactor properly sealed, the pressure stayed high and the ethylene refused to react at all. It was only after months of additional work that they key was found: a trace of an oxidising agent had to be present for polymerisation to occur.

It then became apparent that there had been not one, but two, timely and fortuitous accidents. In the original experiment, the benzaldehyde used had not been freshly distilled, and therefore probably contained traces of peroxide that acted as a catalyst ('initiator'); in the successful experiment with ethylene alone, minute traces of oxygen were present in the cylinder of ethylene gas. Only because the ethylene leakage occurred and was continuously replenished was enough oxygen introduced to effect polymerisation.

It is hard to see how the ICI team could have achieved the success they did or carried the day with their management without the benefit of these 'happy accidents'. Fortune even smiled once more when the first efforts were made to find a use for the new polymer. As Swallow has said: 'It would have been difficult at that time to have carried out a "market survey" which would have led to any profitable conclusion, but here again the role of chance played a part.' ICI happened to have just the right people and contacts to see and pursue a possible use for polyethylene as an insulating material for submarine cables. This first application finally proved unsatisfactory, but it was a fruitful failure, because the experience and the head start on production enabled polyethylene to be nominated and then supplied for the critical job of insulating radar cables. Based on the estimated requirements for submarine cables, a plant had been started, and it came on-stream in September 1939, on the same day that the Germans invaded Poland.

The world knows that radar was a key to victory in the Battle of Britain and in the anti-submarine campaign, and thus a major contributor to the final success of the Allied cause. According to Sir Robert Watson Watt, the English discoverer of radar:

> The availability of polythene (polyethylene) transformed the design, production, installation and maintenance of airborne radar from the almost insoluble to the comfortably manageable .... And so polythene played an indispensable part in the long series of victories in the air, on the sea and land, which were made possible by radar.

We shall not pursue further the innumerable technical and commercial developments by which high-pressure polyethylene became a multibillion pound product and a household word around the world. We should note, however, that we now know that both the desirable properties of this first member of the polyolefin family (softness, flexibility, toughness, easy fabrication) and its property limitations (too soft, too weak, too low melting for some applications) arise directly from the fact that its molecular chains are not straight, but highly branched, and therefore do not fit together very well. Hence there is only a limited amount of reinforcing 'crystallite' formation*. And so 'high-pressure

---

**See* diagram on p. 7.

polyethylene', in the course of becoming the world's foremost plastic, created a potential demand for the products of improved hardness and strength that were to come, 20 years later, from the discoveries of the 'chain straighteners'.

Missed opportunities

In 1937, Hall and Nash, in England, noted that aluminium alkyls are formed during the polymerisation of ethylene with aluminium chloride. In post-Ziegler retrospect, this result appears more significant than it did at the time. Ziegler's observation was the reverse side of the coin; he noted the polymerisation of ethylene during the formation of aluminium alkyls. The other difference is that Ziegler recognised the significance of his result and followed it up vigorously.

Even in industrial research laboratories experienced in polymers, these vital elements of recognition and follow-up were missing on critical occasions. According to Frank Mayo, a key contributor to free-radical chemistry and radical-initiated polymerisation systems, chemists at the U.S. Rubber Company research laboratory made a crystalline polyethylene in the 1940s by a low-temperature, low-pressure reaction (peroxide catalysed!) and sent it to their application group for evaluation. They were told that it was not of practical interest because its 'processability' was so different from that of the familiar high-pressure polyethylene.

The duPont Company itself, pioneer, after ICI, in high-pressure polyethylene, may not have recognised all of its opportunities. DuPont patents, filed as early as 1938, disclose combinations of metal–organic compounds with transition metals that make at least partially crystalline polyethylene. DuPont workers were also able to make high-density polyethylene in a free-radical process operated at really enormous pressures, and although this did not lead to a practical process, it gave duPont a patent and a bargaining position in later licensing negotiations.

As a matter of fact, even the Fischer–Tropsch reaction (catalytic reduction of carbon monoxide with hydrogen under high pressure) is claimed to be capable, under very special conditions, of making high-molecular-weight and high-melting polyethylene. This claim has an ironical twist, since the Fischer–Tropsch reaction originated at the same institute where, under Karl Ziegler as director, low-pressure polyethylene was discovered decades later.

For one reason or another, none of these earlier alternative routes was ever developed into a practical commercial process, nor did the inventors cite regular chain structure as an explanation of the unusual properties of the polymers they obtained.

Another predecessor of Ziegler's came even closer, both chemically and geographically. One of his contemporaries in Germany, M. Fischer, working in the laboratories of the Badische Aniline und Sodafabrik, was trying to improve the well known process whereby ethylene could be converted by the action of aluminium chloride into low-molecular-weight liquid polymers that were of interest as synthetic lubricating oils. He modified the process by adding titanium tetrachloride and aluminium powder (to react with the hydrochloric acid liberated in the reaction). The product was a mixture; half of it liquid and half a white powder. These results, obtained in 1943, were disclosed in a patent assigned to BASF and issued in 1953.

Ziegler's catalyst is a combination of titanium chloride and aluminium alkyl, and it has been argued* that both of these ingredients were probably present in Fischer's reaction mixture (assuming *in situ* formation of the alkyl). Other experts deny this possibility. However that may be, and however crude his product, it would appear that Fischer had a good lead and almost as intriguing a preliminary result as the first crude polyethylene preparation in Ziegler's laboratory. Certainly, his reaction did not require the enormous pressure of the ICI process. Yet, Fischer's discovery led to a patent and nothing more, while Ziegler's created a new industry and fathered a new branch of science and technology.

What made the difference? Undoubtedly, the disruptions of the war years can be blamed in part. But even postwar, it is doubtful that the almost unique combination of internal strengths and favourable external circumstances that Ziegler exploited so successfully would have been found within the BASF organisation and the economic and business communities *sans* Ziegler.

Another famous university figure also scored a near miss, and at an even earlier date. This was Prof. C. S. Marvel, the highly productive (witness his nickname, 'Speed') and respected organic chemist at the University of Illinois (now at the University of Arizona). In 1929, he assigned one of his graduate students (M. Friedrich) to an attempt to prepare organic compounds of arsenic. In one experiment using lithium alkyl in the presence of an arsenic (arsonium) compound, they expected ethylene gas to be liberated but found practically none. In the words of their 1930 publication†: 'This was the cause of considerable concern until it was found that ethylene was rather rapidly polymerised by a solution of a lithium alkyl to give non-gaseous products.'

---

*In patent cases still in litigation over two decades later.

†Friedrich, M. E. P. and Marvel, C. S., *J. Am. Chem. Soc.*, 52, 376 (1930).

Marvel has a well earned reputation as a practitioner and a superb teacher of the arts of the skilled experimenter and observer, and he later became a world figure in polymerisation research. Nevertheless, on that early occasion he did not follow up what today looks like an intriguing lead. His paper simply states: 'This solid has not been studied further.'

Actually, it was investigated further, although not by Marvel. DuPont, for whom he was a consultant, tried out the reaction but did not see any commercial possibilities. Their negative conclusion probably reflects both the inherent difficulties in the process and their limited expertise and interest in high polymers at that time (Wallace Carothers' polyamide precursor to nylon was then a new laboratory curiosity, and high-pressure polyethylene had not yet been discovered). It would appear that this is another case of a promising lead being discovered ahead of the most propitious time. In consequence, the Year of the Polyolefins did not arrive for another two decades.

However long the shadows cast by all these prior events, they neither predated Ziegler's own interest in metal–organic compounds nor prevented his obtaining patents throughout the world on the catalysts and processes developed in his laboratory: catalysts that worked much better under much milder conditions than any previously described. In fact, Ziegler's accomplishment, far from being diminished by his predecessors' actions, is made more impressive by comparison with them. The man who lands his fish while half-a-dozen others are losing theirs has a just claim of being an exceptional fisherman.

Natta's predecessors

There were as many path blazers for Giulio Natta as for Karl Ziegler, who must be counted foremost among them. This despite the fact that no-one (with the possible exception of Ziegler) made highly crystalline polypropylene or Natta's other novel polymers before he did.

The basic idea that stereoisomerism could, in theory, be exhibited by a long-chain polymer cannot be said to have originated with any one individual, since it undoubtedly occurred to a great many chemists. After all, the only requirement for existence of stereoisomers ('mirror images') in an organic molecule is that four different groups be attached to the same carbon atom, making it asymmetric. Since the attachment points are at the tetrahedral angles, two different structures are then possible, which differ only in being mirror images of each other:

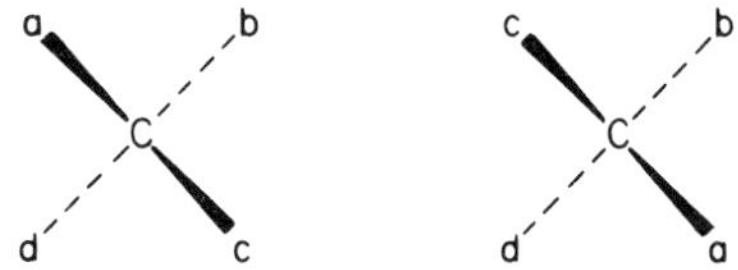

L-isomer ('left-handed') *D*-isomer ('right-handed')

In these diagrams the bonds shown as triangles should be visualised as lying above the plane of the paper; those shown as dotted lines lying below it.

In an organic polymer made from a substituted ethylene, every other carbon atom is asymmetric, if one considers the two sections of the polymer chain on either side as two different groups:

L-isomer *D*-isomer

In an ordinary polymer, one would expect (and find) that any given atom might be in either the *D* or *L* configuration, and therefore that one would have a completely random sequence along the chain:

. . .-D–L–D–L–D–D–L–L–L–D–L–L–L–D–D–L–D–L–L–L–D–. . .

But if, somehow, all asymmetric carbon atoms were in a single configuration, a new degree of order would be introduced: 'stereoregularity':

. . .-D–D–D–D–D–D–D–D–D–. . . or . . .-L–L–L–L–L–L–L–L–. . .

This obvious possibility was long ignored or dismissed for two reasons. First, a polymer chain of any significant length would presumably have an equal number of *D* and *L* atoms in random sequence, and no means of achieving an asymmetric polymerisation had been visualised*.

Secondly, even if such a polymer could be made, the chain segments on either side would differ only in length (say, for example, 850 as against 720 carbon atoms), not in composition, so the effects would probably be too small to be of practical significance, or even to be detected by ordinary means (e.g. optical activity). The possible profound effects on, and results of, close 'fitting' between chains was generally overlooked. It therefore hardly seemed reasonable to spend research time and thought on an effect that probably did not exist or, even if it did, probably could not be detected.

Fortunately, there are some unreasonable people, even (particularly?) among scientists. One of them was Prof. Hermann Staudinger, whom we have already introduced (Ch. 1) as the 'grandfather' of modern polymer theory. In 1932 he published the first serious discussion of the possibility of stereoregular structure in polymers. However, his suggestion had a negative form: that the failure of a particular polymer (polyvinyl bromide) to crystallise might be due to random *D* and *L* configurations along the chain. But he had recognised the one place one should look to find an important effect of a stereoregular structure—the ease of crystallisation. The chains of an 'all-*D*' polymer would have a much more regular structure than a random polymer, should fit together more easily, and should therefore form microcrystals more readily†:

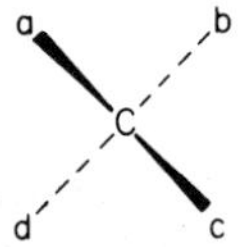

L-isomer ('left-handed')

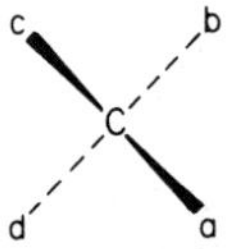

D-isomer ('right-handed')

An American physical chemist, Prof. Maurice Huggins, was the first to invoke steric order to explain differences in behaviour between two chemically identical polymers. Huggins is a brilliant theorist who

*Even the special case of polymerising just one form of a monomer that itself has two stereoisomers (e.g. propylene oxide) was not treated successfully until relatively recently.

†Many different aids to visualising the greater regularity of the stereo-controlled structure have been proposed by various writers. One favoured by a former colleague is one of the simplest: 'Just think of a long school or office corridor with all of the rest rooms on one side of the hallway.'

has made original contributions on many different subjects. In the process, he has earned a reputation as a 'radical' because he has repeatedly been ahead of others in suggesting fundamental explanations for new phenomena: explanations which, more often than not, prove eventually to be correct. In the early 1940s, he proposed an original hypothesis which appears more significant in retrospect than it did at the time. The instigation was a publication by another original thinker in polymers, Prof. Herman Mark.

An articulate, energetic* and charming Austrian chemist, Mark is now a household word in the house of polymers, best known for his indefatiguable activities in communicating information and ideas within the scientific community. When, as still a young man, Mark had been director of polymer research for the giant German chemical firm, I. G. Farbenindustrie, there had flowed from that laboratory a remarkable series of new synthetic polymers, including the first commercial polystyrene. The resultant embarrassing discovery that polystyrene made in a big reactor differed seriously from that made in laboratory apparatus left Mark with a continuing interest in the properties of polymers made at different temperatures. He later published a detailed study which described those differences in terms of a theoretical equation developed by Prof. Huggins, and suggested that the difference might be due to different degrees of branching in the polymer chain.

However, when Huggins himself read Mark's paper, he thought there might be a better explanation, based on differences in steric order. He had a chance to test this idea on other scientists at the next Gibson Island Research Conference. Huggins recalls that while relaxing on the beach between morning and evening sessions, he advanced the concept to several recumbent professional colleagues that the differences noted by Mark had occurred, not because higher polymerisation temperature increased branching, but because it reduced steric regularity.

Since no-one disagreed with this hypothesis (whether from true concurrence or lassitude is not evident), Huggins was encouraged to publish it, stating quite unequivocally that polymers made at low temperature

---

*Mark's energy and the adage about ill winds are both illustrated by the profitable use he made of enforced idleness as a youth in World War I. Taken prisoner in Italy, he was allowed to request reading materials, and said he would like to learn chemistry. Offered chemistry textbooks in German, Mark asked for Italian texts instead, so he could learn both chemistry and Italian at the same time!

should have a more regular structure, 'such as one in which all of the R groups would be on the same side of the plane . . . or one in which the R groups alternated from one side to the other'. He thus anticipated the two structural types that Natta was latter to demonstrate experimentally and to christen isotactic and syndiotactic. Huggins's paper added: 'Experiments are in progress here to test this hypothesis.'*

But Huggins was better at predicting what molecules would do than what people would do. The powerful theoretician had limited experimental manpower at his command, but he expected to persuade others at the Eastman Kodak Company laboratories, where he was then a Research Associate, that the experimental proof should be sought. Unfortunately, he was unable to convince management that the work would be of any potential importance. Hence, the 'experiments in progress' never progressed beyond the planning stage. Nevertheless, Huggins remained (and remains) convinced that his hypothesis was valid, even for polystyrene made by conventional methods.

Ironically, something over a decade later, when stereospecific polymerisation was a proven fact and a hot new field, Huggins made a new proposal to publish some new and pertinent experimental work in which he had collaborated, only to be turned down again, this time because the work was *too* important (in relation to Eastman's commercial position)!

Another dean of polymer theoreticians who recognised the theoretical possibility of stereoisomerism in polymers was Dr Paul Flory. Flory is well known for his careful research and profound reasoning in many branches of polymer science over many years, and his book, '*Principles of Polymer Chemistry*', is a classic. His many honours include the Priestly Medal, top honour of the American Chemical Society, the National Academy of Science Medal, the Nobel Prize in Chemistry (1974), and the Perkin Medal of the Society of the Chemical Industry, awarded in 1977. The garnering of these high honours by a polymer scientist has been heralded as presaging the end of the myopia that has kept many leading universities in the U.S. from recognising polymer science as a major field and discipline both for education and basic research.

Flory had left the academic field fairly early for a career in industrial research at duPont, Standard Oil Company of New Jersey, and Goodyear Tire and Rubber Company, but his experiences finally convinced him that industrial research was not his proper métier.

---

*Huggins, M., *J. Am. Chem. Soc.*, 66, 1991 (1944)

In those days, Flory was not known for his diplomacy or patience in dealing with lesser intellects, and he could not get on at all with Goodyear's director of research. The director promulgated many fussy restrictions on working hours, opening windows, closing drapes, etc., in the laboratory which Flory considered intolerable pettifogging and encouraged his group to ignore as a matter of principle. He left Goodyear in 1948 to return to more academic environments (Cornell, Mellon Institute, Stanford), where his highly fundamental and theoretical approach would be better appreciated*.

Flory's recognition of the possibilities for polymeric stereoisomerism, like Huggins', had been a negative one, in that he saw random *D* and *L* sequences as the most probable structures and the reason why crystallinity or optical activity were 'almost unknown' among polymers with asymmetric carbon atoms. Although he cited Schildknecht's polyvinyl ethers (discussed below) as the 'sole possible exception' and characterised his hypothesis as 'plausible', Flory did not himself follow up or actively support efforts to confirm it. It has been suggested that his preoccupation with mathematical modelling may have diminished his enthusiasm for inductive research, where the experimental observation precedes the theory.

Despite Huggins, Flory and others, it remained, as it always must, for experimentalists to explore and claim the country the theoreticians had viewed from their mountaintops.

One group of experimentalists in the duPont laboratories (Anderson, Merckling and co-workers), according to patents filed in 1954, had found catalyst systems involving titanium chlorides and lithium aluminium alkyl that gave partly crystalline polymers from propylene. However, just as we saw in the case of polyethylene, they did not attribute their results to stereoregularity, nor did duPont attempt to capitalise on their findings, at least not until they became involved in litigation with Montecatini over Natta's patents.

Calvin Schildknecht

The man who may be said to have been in the first line of advance in the direct attack on the problem of stereospecific polymerisation is Prof. Calvin Schildknecht. Like most members of even the second and third

---

*Asked why he left industry, Flory replied: 'I got tired of casting synthetic pearls before real swine'. If there is an Acid Aphorism Award, he must be a candidate for that also.

generations of leaders in polymer chemistry, his formal degree was not in that specific discipline, but in physical organic chemistry. At Johns Hopkins, he was influenced by professors such as Emil Ott (later director of research for Hercules) and Maurice Huggins, the pioneer thinker about polymers whom we met above, so it is not surprising that Calvin Schildknecht entered the polymer field. In his first industrial research position, at duPont, he came under the further influence of Carl Marvel and Herman Mark (who were duPont consultants) and of a duPont secretary, Althea Schneider, whom he married.

All paths in polymers, if pursued forward and backward, seem to lead at some point to Germany*. Here the trail runs back to a much-publicised German chemist, Dr Walter Reppe, who carried acetylene chemistry so far forwards that a new branch of the chemical industry arose from the base known as 'Reppe Chemistry'. It dealt with small organic molecules, not polymers. But among them was a family of unsaturated ethers (the vinyl ethers) that were capable of being polymerised to high polymers.

Reppe's work was done mostly between the two World Wars, under the aegis of Badische Aniline und Sodafabrik, one of Germany's major chemical firms. This company had an American subsidiary, General Aniline and Film Company, which naturally had access to the information about Reppe's work. World War II of course terminated contacts with the parent company and, at the end of the war and a tortuous series of legal battles and corporate manoeuvres, the divorce decree in the case of General Aniline *vs* Badische Aniline was made final†.

Calvin Schildknecht, who had left duPont and joined General Aniline as a research chemist in 1943 (the same year that Karl Ziegler came to the Max Planck Institute in Mülheim, Germay), was soon involved with that company's efforts to develop new products from Reppe chemistry for American needs. The work included making polymers from vinyl ethers. The research team included (besides Schildknecht) A. O. Zoss, an organic chemistry Ph.D. from Notre Dame, and several members

---

*Schildknecht himself is of German and Swiss ancestry. Somewhat appropriately, 'Schildknecht', in German, translates literally as 'shield bearer'.

†General Aniline and Film subsequently followed the fad among American companies to obfuscate their origins and assist the Stock Exchange tote board by shrinking their names to unexplained initials. Thus, General Aniline is now simply GAF.

of the physics department: Siegfried Gross, trained in x-ray spectroscopy at the University of Illinois, Joseph Lambert from the University of Vienna (where Herman Mark had been one of his teachers!) and Hugh Davidson from Lehigh University.

In 1945, the polymerisation of vinyl isobutyl ether was being studied on pilot plant scale. One day in September, a batch of polymer came out in a highly unusual form. Schildknecht and his co-workers set out to find out what had happened and then how to control it.

In April 1947, Schildknecht described to the American Chemical Society* the peculiar polymerisation behaviour of vinyl isobutyl ether, and a few months later he repeated the description at the Gordon Conference on high polymers.

The Gordon Conferences, the successor to the Gibson Island Conferences mentioned above, are a highly regarded series of summer seminars at which American scientists and technical people from academies and industry, plus invited foreign visitors, gather to hear and discuss lectures on new findings in many different fields. Informality and free discussion are fostered by prohibiting publication of papers presented there, but the size and calibre of their audience ensures rapid dissemination of any exciting content.

At the Gordon Conference on high polymers at New London, N.H., in June 1947, Calvin Schildknecht discussed and demonstrated two different modes of polymerising vinyl isobutyl ether. As catalyst, he used boron trifluoride (BF_3), a well known, vigorous agent for polymerising many monomers. First, he added a solution of BF_3 to the ether all at one time, causing a violent, almost explosive reaction that immediately produced a voluminous mass of solid polymer that was tacky and rubbery. He then repeated the experiment but with the catalyst added very gradually and the system kept chilled with dry ice (the BF_3 catalyst was also 'tamed' by complexing with another ether).

The result was now less spectacular but more intriguing: a slow growth of solid polymer that was not tacky or rubbery and was much harder than the first. It could be shown to be highly crystalline, while the rubbery polymer was completely non-crystalline.

Why the difference between these two polymers, identical in chemical composition and made from the same materials? Schildknecht suggested, reasonably, that the crystalline polymer must have a more regular chain structure that enabled adjacent chain segments to fit together in compact bundles or layers.

---

*At the Atlantic City, N.J., meeting of 16 April 1947.

It would have been safe for him to suggest further that this regularity resulted from diminished chain branching (by analogy with what had already been postulated for other known polymers), but Schildknecht instead made the bold proposal that it arose from a regularity of spatial configuration, as would occur if the carbon atoms in a given molecule were nearly all in either the *D* or *L* configuration (i.e. all side groups on one side of the chain), or in regular alternation (*DLDLDLDLDLDL*). These possible structures would be represented as follows (letting R stand for an isobutyl group):

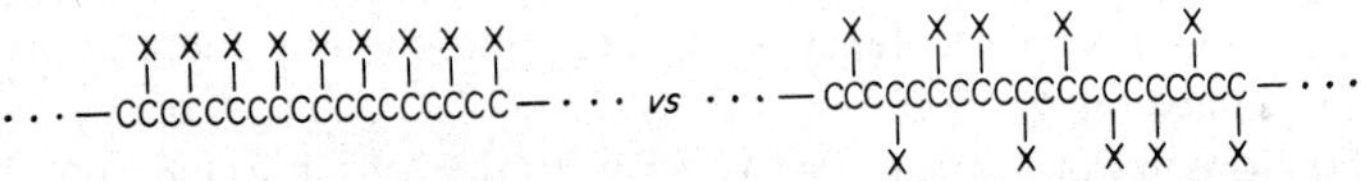

The audience was impressed by the demonstration but not by the explanation. At a subsequent lecture, in fact, there was so much hard questioning and scepticism that Herman Mark, who was presiding, had to intervene to re-establish decorum.

But even Mark, whose receptivity toward new ideas was well known and would be demonstrated again in the case of Karl Ziegler, did not give Schildknecht's hypothesis instant acceptance. I hunted him up in the bar at New London on the evening after Schildknecht's talk to ask him what he thought of it. With his unfailing polite good nature, Mark smiled and said: 'It's a very interesting phenomenon and Dr Schildknecht has a very ingenious idea. But I really think there must be some other explanation.'

Typically, however, this did not deter Mark from cooperating with Schildknecht in measuring the molecular weights and viscosities of the polymers. Later on, Mark played his usual generous role by publicising Schildknecht's work and helping to see that his contributions were given due recognition.

Schildknecht had previously received encouragement from Maurice Huggins when they met at the Gibson Island Conference a year earlier. Calvin had made brief mention of his work and theories with regard to polyvinyl ethers, and Huggins had shown immediate interest. Walking on the beach afterwards, Huggins eagerly drew formulae in the sand to illustrate possible stereoisomers for polyvinyl alcohol (PVA), closely analogous to the polyvinyl ethers. Schildknecht irreverently reminded his former professor that his formulae, as written, were not PVA but SiO_2 (silica sand!). But he was serious enough about the encouragement he derived to pursue detailed studies on the polyethers, which he reported the following year.

In a 1949 publication* (five years before Natta), he diagrammed and discussed the various possible spatial arrangements for a polymer chain of asymmetric carbon atoms; however, he favoured the regularly alternating structure (which Natta was to christen later, 'syndiotactic'), as the most probable for the crystalline polyvinyl ethers.

'It seems less probable', he wrote, 'that all of the OR groups in a macromolecule could occur on the same side of the chain and this condition is not represented.' However, Paul Flory pointed out in 1953 that the 'all-*D*' structure with all side groups on one side might be the favoured configuration. In Natta's first publication on stereoregular polymers, he concluded that this structure ('isotactic') was the correct one and could be established for polyethers by crystallographic measurements.

Despite the initial publicity and scattered encouragement, Schildknecht's work did not ignite the imaginations of either the scientific or industrial communities. He kept the subject alive in subsequent publications and discussions, but his novel polymers continued to be regarded more as 'sports' than as the first of an important new breed. In contrast to the bandwagon fever that was to greet and almost overwhelm Ziegler and Natta, Schildknecht's peers, and General Aniline's competitors, were generally content to leave it to the originators to see what they could do with the discovery.

The company never succeeded in developing vinyl ether polymers into important commercial products, and Schildknecht never came up with the clear elucidation and clinching evidence that would unequivocally establish the reality and importance of stereospecific polymerisation—evidence that Natta so brilliantly provided a few years later.

Schildknecht left General Aniline in 1951 and industrial research in 1953 to return to teaching, first at Stevens Institute and later at Gettysberg College. Belated recognition came his way in the aftermath of Natta's prize-winning accomplishments, as we shall see in later chapters.

Many factors are responsible for the limited impact of Schildknecht's publications compared to the worldwide acclaim and commercial consequences we shall find following hard on Natta's discovery and christen-

---

*Schildknecht, C., *et al.*, 'Isomerism in Vinyl and Related Polymers', *Ind. & Engng Chem.*, 41, 1998 (1949). The same subject is discussed in his book: *Vinyl and Related Polymers*, Wiley, New York (1952).

ing of isotactic polypropylene. These factors will be reviewed in our closing chapter on 'interpretations'. Unquestionably, one of the important differences is the fact that Natta benefited from the information and attention generated by the discoverer of 'Ziegler polyethylene'– Prof. Karl Ziegler. It is to that remarkable man and his work that we shall turn our attention in the next chapter.

3 Karl Ziegler and the Max Planck Institute

My only motivation has always been just to do what was fun.
—Karl Ziegler

Ziegler, the man

It is often interesting, and sometimes instructive, to trace a distinguished career back to its beginnings—to spot the decisive event, the occasion when Destiny said clearly, 'come this way!' By Karl Ziegler's own account, this turning point came very early for him.

The man who was to discover the new field of chemistry that bears his name and would win for him the Nobel Prize was not born to science, but the Muse that was to crown him as a man first claimed him while he was still a boy. His father was a clergyman, and Karl, the youngest son, was born in the parsonage at Helsa, near Kassel, in 1898. In 1910, when Karl was twelve, the family moved to Marburg, where his father had studied theology. Renewing his contacts with Marburg University, the minister brought many members of the faculty, not just the theologians, to his home. In this way, his son became acquainted with the persons and thoughts of eminent scholars, including scientists.

But the event that determined the direction of Karl Ziegler's life interest had occurred earlier; at the impressionable age of eleven he had encountered a 'primitive' textbook of physics. Immediately fascinated by the world of science thus discovered, young Karl did much independent reading, which he continued through his student days at the Gymnasium. In addition, as so many other future chemists have done in their youth, he set up a little 'home laboratory' in which he could do his own experiments.

There are several slight threads that connect Ziegler's youth with Marburg University and even with the Kaiser Wilhelm Institute (later, the Max Planck Institute), although too tenuously to pass for predestination. Karl's elder brother had died at an early age of diphtheria and, by coincidence, the 'conqueror' of that disease (i.e. the discoverer of diphtheria serum), Dr Emil von Behring, was the director of the Hygienic Institute at Marburg.

Not surprisingly, the illustrious physician was held in particular regard by the elder Ziegler and, through his influence, by the children. Although Karl never met von Behring, he was confirmed with one of his sons. Moreover, Karl indirectly 'basked in his radiance', as he put it, when he won the prize that von Behring had established to be awarded to an outstanding senior each year—a 'stipend' to visit the German museum at Munich. This modest prize, the first of his career, affected Ziegler more than most of his many later awards. He was greatly impressed by the museum's collections and program, but perhaps even more by the opportunity the trip gave him for his first experience in mountain climbing—an ascent of the Zugspitz. This was the beginning of a lifelong interest in mountaineering.

Another eminent scientist at Marburg, Alfred von Harnack, was involved in the founding of the Kaiser Wilhelm Institute. Ziegler's father once remarked to him that that Institute 'might someday be something for you.' Seeming prescience aside, the remark is of interest as evidence that here, at least, was one clergyman father who had no objection to his son's choice of science as a profession. However, when one of Karl's schoolmates (Hans Broche) went off to work at the Kaiser Wilhelm Institute at age twenty, young Ziegler, showing complete lack of prescience, said that he could not imagine doing such a thing!

Thanks to his quick mind, independent reading and home experimentation, Ziegler was able to answer questions and demonstrate analyses that showed he had already mastered the first-year material at Marburg University, and thus to persuade the faculty to admit him at the third-semester level. Once in the university, he benefited greatly from the extensive personal attention afforded him by his professors. One of them, who had already advised Ziegler to stay on for postgraduate work, nevertheless introduced him to a key man in the big chemical company, Hoechst, presumably paving the way for a later job offer. But this was wartime (1918), and the student was called to the army instead. Soldiering was one activity for which Ziegler disclaimed any talent or inclination and one from which, fortunately, the end of the war soon released him.

Returning to the university, Ziegler completed his studies and received his doctorate in chemistry in 1920, after a total of only three and a half years of study, and four months before his twenty-second birthday. This was something of a record, as was his appointment as a lecturer at the university in 1923 after two years of postdoctoral research. He had obviously applied himself with great diligence to study and research, but not to the exclusion of a young man's normal other interests. He married a Marburg girl in 1921 and in 1922 their first child was born.

There had been those who thought twenty-five was much too young to make even a precocious young man a 'Privatdocent', and there were those who, observing young Ziegler going about Marburg in careless attire (baggy knickers and sloppy leggings), had remarked that he 'could never hope' to become a 'Herr Professor'. Time soon disposed of their doubts as well as of his excessive youthfulness.

It also disposed of Ziegler himself, as far as Marburg was concerned. In 1925 began that periodic migration that marked the advancing academic then, as now. He went first to Frankfurt, then to Heidelberg and, finally, to Halle, where he became head of the Department of Chemistry. Throughout these moves, he maintained the research interest in metal-organic compounds that had occupied him first at Marburg and that was to culminate much later in his discovery of 'Ziegler Chemistry' and 'Ziegler Polyethylene'.

But this was by no means his only line of investigation. The moon pales at sunrise, and Ziegler's numerous other research achievements are dimmed by the brilliance of his polyethylene discovery. However, they are both known and noteworthy, and although not part of this story, some will be listed here as evidence that his most famous achievement was neither his first nor his last. Chemists will recall, or will learn on consulting the literature, that significant contributions have been made by Karl Ziegler on such diverse subjects as the following:

free-radical chemistry,
large-ring compounds,
synthesis of cantharidin (active principle of 'Spanish Fly'),
polymerisation of butadiene (enters our story later).

The first-named subject is one that invalidates characterisation of Ziegler as a classical organic chemist. While not the 'father' of free-radical chemistry, he was certainly one of the leading early contributors.

That work caused one of his contemporaries to describe Ziegler as 'one of the first physical–organic chemists'. It is also the basis of a story Ziegler has told on himself. On his first visit to England, at a time when Hitler was on the rise and relationships were already rather touchy, he was asked by the English customs agent the purpose of his visit. Ziegler replied that it was to give a lecture.

'What is the subject of your lecture?'

'Free radicals'.

Understandably, this disconcerting reply caused considerable stir and a lengthy delay until the authorities could be convinced that chemical free radicals are apolitical.

Ziegler comes to the Max Planck Institute for Coal Research

Germany has a long tradition of esteem for scientific research, derived from and, in turn, fostering, an outstanding record of achievement by her scientists. This high regard is reflected in the elitism granted to eminent scientists. It has been remarked that an encounter with a professor is apt to produce the same response in America and in Germany—a finger to the forehead. But the connotations are exactly opposite: in America, a gesture of derision; in Germany, a respectful salute.

The esteem, the elitism and the achievements are all exemplified in the story of the Max Planck Institute and Karl Ziegler. Originating with the Kaiser's sponsorship in 1910, the Kaiser Wilhelm Institut für Chemie opened in 1912 in Berlin and was joined in succeeding years by others devoted to specific fields of science. All were connected through the Kaiser Wilhelm Gesellschaft (Society). After World War I, a change in image was called for, and the name of the patron of German science was dropped in favour of one of its most illustrious practitioners, the father of the quantum theory. Thus the Kaiser Wilhelm Institut became the Max Planck Institut.

The Max Planck Institut für Kohlenforschung (Coal Research) was created with the support of a consortium of German coal companies, who provided the financing with the expectation that the institute would do research related to and, hopefully beneficial to, the coal industry. Located conveniently at Mülheim a.d.Ruhr, it was opened in 1914 under the able, if autocratic, leadership of its distinguished first director, the redoubtable Dr Franz Fischer (cousin of Emil Fischer, the famous pioneer in the chemistry of natural products: carbohydrates, amino-acids, proteins, etc.).

The hoped-for benefits were realised in due course and full measure.

Most notable of many achievements was the now-famous Fischer–Tropsch process for making synthetic hydrocarbons, particularly synthetic gasoline. Fischer died during World War II, and a successor had to be found to carry on the work in those trying times.

The governing body of the institute now included representatives from both industry and the prestigious Max Planck Society. By this time (1943), Karl Ziegler had been head of the chemistry department at Halle for some seven years, and his researches and publications had already won for him an outstanding reputation in the scientific community. The Max Planck representatives made, and the industry representatives accepted, Ziegler's nomination to the vacant director's post; but *choosing* and *getting* proved to be two different things. At the first contact, Ziegler told his visitors: 'I don't know anything about coal, I never did anything with coal in my whole life, and I don't want to. It's very nice of you to visit me, but I'm the wrong man for you'.

Despite this rebuff, the scientists still insisted that Ziegler was their man. What they wanted above all was the highest possible scientific ability, and they were sure that it would be better to have a first-rate independent director than a second-rate sycophant. The industry members, on the other hand, were dubious. Business men like to have control over the spending of their money and some assurance of a return on it, and the coal company executives were no exception. But they were finally persuaded to settle for the very general assurance that, while organic chemistry was not coal chemistry, there was a good chance that organic chemistry would produce something quite useful.

Ziegler had also made it clear that he would not accept any outside control in either choice or pursuit of research goals. Although his playing so hard to get suggests a certain amount of gamesmanship, he was probably playing it pretty straight; he simply did not want to take a position that would restrict his freedom in research, and since he already had a scientific reputation and a research institute at Halle, he felt no need to compromise. In later years he wrote, with obvious pride and hyperbole, both pardonable, that the research that had proved so fruitful was to him simply 'fun' (Späss), that his only 'drivespring' (Triebfeder) was the pleasure he took in what he was doing, and that he had never been forced to do anything other than to indulge his scientific curiosity.

In the end, Ziegler got everything he asked for in his contract with the Max Planck Institute. That remarkable document not only guaranteed him autonomy in directing research and freedom of publication, but also granted him personally the rights to any useful inventions that fell outside the field of interest of the sponsoring coal companies. His sponsors would have had no cause to regret the first of these provisions had it not been for the last. They obviously could not foresee the rivers of money that would rise as a result of Ziegler's work and flow to him as a result of his contract. Only after that flow had made him a very wealthy man did they take steps to modify their contract with him, and even then only to divert money to defraying the operating costs of the Institute*.

The foreign observer is bemused, even amazed, to reflect that these arrangements were negotiated in wartime, that in a nation whose every resource was committed to a conflict that would soon decide between survival and disaster, its industrialists could find the time, money and will to concern themselves so seriously with the long-range future of a long-range research facility. It is even more surprising that they could so completely free it from any compulsion to support directly the national life-and-death struggle of the moment.

Certainly, this was not blind devotion to pure science for its own sake; rather, it must have reflected optimism about the war's ultimate outcome, plus a determination to continue a winning game based on the past record of the Institute. Under Fischer, the research program had yielded synthetic gasoline and numerous other goodies; might not similar breakthroughs be expected in future; even in the near future?†

This demonstration of faith in 'non-targeted' research, made under the most trying circumstances imaginable, stands in stark and brave contrast to the view currently in vogue in industry that any money spent on research must be tightly controlled by business managers to

---

*Ziegler himself made a gracious ceremony out of dedicating substantial funds to his Institute. On the occasion of his twenty-fifth anniversary as director and his own seventieth birthday, he announced the founding of the 'Ziegler Fund' with an endowment of forty million marks (roughly ten million dollars) to be used in support of the Institute's research.

†Alternatively, it has been suggested that the German industrialists were already looking ahead to the postwar era, realising that, even if the war were lost, industry would be rebuilt and would then need the new products that research might discover. Foreseen or not, that was the outcome.

ensure 'gearing' to the 'real' needs of the business. Such controls of course also ensure that no real innovation is apt to be made or recognised; under them, Ziegler could never have worked with metal alkyls or discovered linear polyethylene. It was his freedom from dominance by 'commercial' thinking that enabled him to make a landmark discovery of such commercial importance. His sponsors' act of faith was thereby justified; ironically, the wording of their contract with him denied them its full benefits.

Ziegler's appointment to the Max Planck Institute soon proved to have an unsuspected additional advantage for him. When he accepted the directorship at Mülheim, he did not move to that city, but retained both his chair at the University and his home in Halle. This seemed a wise precaution under the difficult and dangerous wartime conditions, for Mülheim, in the heart of the Ruhr, was under constant threat of air attack (so much so that the Institute's library, considered an irreplaceable asset, was moved to the basement for safe keeping).

For a time, Ziegler had himself driven to Cologne, rather than Mülheim, and met with his staff in the Dom Hotel, under the shadow of the great cathedral. The divine protection thus invoked was indirect; it was hoped that the bombers would avoid the cathedral and thus its immediate environs. They did, and both Dom and Dom Hotel escaped destruction, if not damage.

Meanwhile, back at Halle, Ziegler's wife, Maria, was coping with the problem of feeding her family. Part of her solution was to raise chickens, ordinarily an unimaginably undignified activity for the wife of a Herr Doktor Professor Director.

At war's end, Halle found itself in the path of the advancing Russian forces and lost its advantage as a seat of operations. One account has it that on a certain day in 1945, a big U.S. Army truck with a black driver arrived in front of Ziegler's house. He was asked: 'Do you want to go East or West?' He promptly chose West, and had two hours to pack before being driven, family and baggage, to Hanover and, subsequently, to Mülheim.

In the backwash of war, Ziegler was assigned by the Occupation forces to the preparation of a detailed review ('FIAT' report) of the German chemical industry for the benefit of the Allies. He subsequently figured prominently in the rebuilding of the German chemical industry and held many positions of responsibility and honour in scientific and technical societies. When Marshall Plan money became available, he applied some of it to the purchase of laboratory equipment to modernise the Max Planck Institute's facilities. And when the occupying powers agreed to the re-establishment of the German Chemical Society (Gesellschaft Deutscher Chemiker) in 1949, Ziegler was elected president.

Ziegler as a research director

When I first met Karl Ziegler, he seemed the very model of the classical German professor and 'Herr Doktor'. This impression was borne out by his physical appearance and manner: the close-cropped, iron-grey hair, the steel-rimmed spectacles, the serious mien, the formal dignity of dress and bearing, and the precise phrasing and self-assurance displayed in conversation.

But there is no complete archetype, as there is no average man. Ziegler was very much an individual; he displayed too much understanding of physical chemistry for a classical 'Organischer', too keen an interest in practical applications for a 'pure' scientist, and too much humanity for the stereotype scientific autocrat.

There is no doubt that, once he took up residence in Mülheim, Ziegler made the Max Planck Institute *his* institute. Those who studied or worked under him there describe a tightly directed and firmly, though kindly, paternalistic environment. The Director's home was an integral part of the Institute building, so that Ziegler could 'commute' to his office simply by walking down the hall. Often his wife, Maria, usually referred to simply as 'Frau Professor', would have to telephone him at midday to remind him to leave his office long enough to come 'home' for lunch, and again at dinner time.

In these days when management theory has been elaborated to a pseudoscience and even modest organisations are multirank pyramids, it is impressive to note that the professional organisation at Mülheim boasted only two levels: (1) Karl Ziegler, and (2) everyone else. This strictly horizontal structure meant that Ziegler had to direct and coordinate virtually all the research himself and had to spend considerable time almost daily with each chemist. And that is precisely what he did. If he were leaving for even a weekend, he would have a special meeting with each of his co-workers to lay out the program to be followed during his absence. The tender flower of personal initiative tended inevitably to wither in this climate, and whenever Ziegler's return was delayed, so was the research program.

There is no doubt about Ziegler's first love. If, as he asserted, he only did things that were 'fun', chemical research must have been the most fun of all. But if Science was his mistress, that did not vitiate his marriage or his home life. The formidable 'Herr Doktor Professor' was also very much a family man*, and one who clearly enjoyed his domestic pleasures. He was fond of music and had an extensive record collection and excellent playback equipment, both of which he would proudly demonstrate to favoured visitors. He and Frau Professor enjoyed gardening and flowers, and had a room in their home converted to a 'garden room', or conservatory, where they could seat themselves and visitors in easy chairs, surrounded by flowering plants, to enjoy music or conversation.

Ziegler also had a strong interest in astronomy; specifically, in eclipses. In later years, he would travel to any part of the globe that was scheduled to experience a total eclipse. The new multistorey laboratory building at Mülheim, built in the 1960s, boasted a small observatory on the roof for Ziegler's use. When I visited his Institute in 1972, Ziegler was off to Easter Island with his grandson, viewing an eclipse (he became ill on that voyage and died the following year).

He was also a 'family man' in a broader sense, for the Mülheim Institute was in many ways an enlargement of his family, and its buildings an enlargement of his home. Quite a few of the young chemists were Ph.D. candidates, by virtue of an arrangement with a nearby university (Aachen) whereby their research could be credited toward their advanced degrees. Ziegler, as their professor, was thus responsible for their academic training as well as for the Institute's program. Together with his wife, he also accepted responsibility for the personal wellbeing of both students and staff.

Frau Professor is fondly remembered by people who worked at the Institute as a likeable but formidable woman. She took great interest in the students and entered into, and helped to organise, the social life of the Institute. She tried to help anyone who had personal problems, outside or inside the Institute and, on occasion, even interceded with her husband on behalf of a member of the staff. Conversely, to incur her displeasure was to invite trouble, for she exerted considerable behind-the-scenes influence.

As a summation, it would be hard to improve on this extempor-

---

*Ziegler had a son (a physicist), a daughter (an M.D.) and numerous grandchildren.

aneous description of the Zieglers, given to me by an 'outsider' who had become a friend and admirer:

> You have this man who is intensely proud, intensely competent as a scientist, but no ivory tower man. A family man, very proud of his son, very proud of his grandson. A man who'd suffered a lot, because he'd built up his research institute at Halle, and all this had been destroyed, and he'd had to come away with only the possessions he could carry in a car and start again at the Max Planck. But always behind him was this understanding, kind wife who was also part of the Institute, because she was not only his wife, she was a mother to all of the people in the Institute. All the senior people and their wives were her children.

But there was nothing soft, lax or casual about the working relationships. If Frau Professor 'mothered' the staff and her husband was something of a father figure, he was also a firm and exacting director. He engendered respect and admiration but not familiarity. He 'laid down the line' which each man was to follow, and from which he would not deviate save at his peril. Ideas could be proposed, but unless accepted by the Director, they were not to be pursued. The student's position is exemplified in the remark of one of them: 'After all, I was his student, so I had nothing to say.'

What was probably Ziegler's greatest failing was the direct result of his great ability. To quote one who knew and respected that ability: 'Karl Zieger had the intellectual arrogance of one who is undisputed master in his own house. He could not escape the feeling that he knew everything about everything in his field.' We shall see later how that overconfidence worked to his disadvantage when others entered the same field.

All in all, Ziegler comes off pretty well in evaluations by his former students and staff members. They remember him as a hard-working and hard-driving, even demanding, but not unfair, boss. He was sure of his goals, confident in his own judgment to the edge of arrogance, even autocratic, but not to an extent that was insufferable or even unusual in that time and that country.

Ziegler disliked administrative detail and left it to others when he felt he could, preferring to spend his own time thinking and talking about research. Yet, he insisted on being involved in final decisions on all matters, technical or otherwise. He also needed staff assistance in keeping peace with the Institute's financial sponsors, and he got it often from Dr Günther Wilke, a senior staff member who became the closest thing the Institute had to an assistant director. Ziegler had continued an aggressive pursuit of his interest in metal-organic compounds* at Mülheim, and on one occasion, one of the sponsors' representatives asked, somewhat petulantly:

'Why do you work on these metal alkyls constantly? We don't see any connection between them and coal research.'

'I don't either,' Ziegler replied, 'but they are much more interesting to work with than coal.'

Obviously, a buffer was needed to minimise the frequency and impact of this sort of encounter, and it fell often to Wilke to be the buffer. He did the job well; the peace was kept and Wilke, who had also demonstrated outstanding scientific ability in his own research, eventually succeeded to the directorship when Ziegler retired (Wilke actually has no more love for administrative detail than Ziegler, and is said to delegate it more effectively).

Dislike for administration did not stem from any disdain for, or lack of appreciation of, the economic significance of research. That Ziegler was highly conscious of the potentials for practical applications is evident in the terms of the contract he negotiated with the Institute, and in the personal time and attention he gave to the patenting and licensing of his discoveries. In contrast to his own lucrative arrangement, staff members were required to sign over all rights to any inventions they might make†, and one man who had somehow missed doing so in advance was required to make the assignment *ex post facto*. When he refused (having survived a Russian prison camp, his powers of resistance were well developed), he was dismissed; this despite a rumoured intercession by Frau Professor and an appeal to the board of the Max Planck Institute.

Ziegler's insistence on the importance of being able to achieve practical quantities of products in any new synthesis was legendary. To an excited young chemist announcing his test-tube result, Ziegler would say: 'Come back when you have made a kilo, and I shall get excited also.'

---

*Best known are the metal alkyls, in which a metal atom is attached to a hydrocarbon radical, commonly the ethyl radical, C_2H_5.

†Despite such assignment, inventors were legally entitled to share in royalties collected by Ziegler, and those whose names appear on patents along with Ziegler's have profited handsomely.

How little he sympathised with the modern tendency to base patents and processes on nothing more than a blip on a chromatogram is evident from the expression, 'a Ziegler ton', used around the Institute to designate a very large amount of material.

The 'growth' (Aufbau) reaction

By the time World War II was underway, Ziegler's research on organic compounds of metals had been underway for almost two decades. The compounds were new, their chemistry novel and the research of high professional calibre, but practical applications proved elusive. For one thing, the metal-organic compounds were hard to synthesise and dangerous to handle. In addition, practical difficulties turned up in the path of conceptually attractive processes. The early work was done with the familiar alkali metals, sodium and potassium; some work had been done with the much rarer light metal, lithium (including its use for polymerising butadiene, of which more later), but had not reached a decisive stage, partly due to the scarcity of lithium.

Fortuitous circumstance, the traditional patron saint of research, intervened to reward diligence with opportunity. Ziegler's institute was made the wartime repository for the country's limited supply of lithium metal, and he found himself sitting on what may have been more lithium than the world had ever seen in one place before. This windfall both stimulated and facilitated resumption of research with lithium alkyls; specifically, pursuing an idea Ziegler had for a cyclic process whereby lithium would link two or more molecules of ethylene together and then conveniently break off as lithium hydride and go back after another ethylene*. But poor solubility of lithium hydride had prevented its reacting as desired. Since that early work, Ziegler had learned of the discovery of a bimetal compound containing lithium and aluminium (lithium aluminium hydride). Since it was more soluble than lithium hydride, he reasoned that its reactivity should be better. He had some made and tested, and it was.

---

*More formally, lithium ethyl, reacted with ethylene, would form lithium butyl or still higher lithium alkyls, which would then dissociate to form an olefin and lithium hydride. The latter might be reacted with additional ethylene to start the process over again. Since the lithium acted as a catalyst and was not consumed, this was a potentially cheap way of converting abundant, low-cost ethylene into more valuable higher olefins, prized as raw materials by the chemical industry.

At this point, a chemist with a strong eye for practical applications might be expected to start thinking about developing a commercial process, based on lithium aluminium hydride, for making valuable larger molecules out of cheaper little ones. But in Karl Ziegler, scientific curiosity triumphed over pragmatism, and desire for the best precluded settling for the merely better. He wanted to know whether the active metal of the pair was now the lithium or the aluminium. This led him to prepare and test aluminium alkyls, and this was the first real breakthrough. The work was done by H. Gellert, a former Halle student.

Aluminium triethyl was willing to add on, not one, but several more ethylene molecules end to end to form higher alkyls. This reaction was promptly christened the 'growth' (Aufbau, literally 'build-up') reaction. Sooner or later, the aluminium atom cast loose the alkyl group (as an olefin) and went back after more ethylene molecules, to repeat the process more or less *ad infinitum.*

How large the newly formed molecules were depended on how many ethylene units managed to link up before the second reaction, called the 'displacement' (Verdrängung) reaction, intervened to break off the chain. This did not always occur at the same point, so the product was a mixture of several different chain lengths. But the really significant feature was that the chains were always straight, i.e. unbranched. The entering ethylene molecule invariably attacked the end, never the side, of the existing chain, so no branches were formed. This was in marked contrast to all other known processes for making ethylene polymers, either the low 'oligomers' or the super-long-chain polyethylene made by the high-pressure process of Imperial Chemical Industries.

Ziegler believed this was a significant finding and was diligent in publishing, patenting and publicising the results internationally. He gave a number of lectures, some public and some to industrial scientists whose companies had invited him as a consultant/lecturer. As so often happens when one scientist is asked to evaluate another's work, his hearers generally missed the main point because they could see the practical difficulties so clearly. At one leading U.S. industrial laboratory*, for example, the director solicited the opinions of those who had heard Ziegler's lecture, asking whether there was anything worth following up in what they had heard. The uniform response was that the process made such a mixture of products that there would be a serious problem of product separation and a low yield of any specific desired chain

---

*Shell Development Co.

length. This disadvantage impressed most people much more than the intriguing linearity of the products.

Even some eminent representatives of the German chemical industry did not escape the occupational hazard of the industrial scientist: looking upon his academic colleague as an intellectual dilettante whose ideas and work are of dubious practical import. Bayer, one of the big German chemical companies, never took a licence from Ziegler, although they had an early opportunity. Part of the reason may have been the attitude exemplified by Bayer's research director, Dr Otto Bayer. He was among a number of leading German chemists present at a dinner which Ziegler attended and at which he was asked about the importance of the new chemistry coming out of the Max Planck Institute. Ziegler, with characteristic dignified assurance, said he was sure that it would stand as an important contribution and would be known in the future as 'Mülheimer Chemie'. Bayer, polishing his reputation for sarcastic wit, remarked that that would be an unfortunate choice of name for an internationally famous process. The problem would be that all Frenchmen would be sure to mispronounce it as 'Müll-eimer' (initial 'h' being silent in French). In German, 'Mülleimer' means 'garbage can'!

Ziegler was not amused. Nor was he deterred; when, several years later, he published his first major paper on polyethylene, it was titled, boldly, 'Das Mülheimer Normaldruck Polyäthylen-Verfahren' (The Mülheim Normal-Pressure Polyethylene Process)*. However, it was but a short while after that until someone else (Giulio Natta) rechristened it and put the phrases, 'Ziegler Chemistry' and 'Ziegler Polyethylene' into the language and the literature.

Fortunately for Ziegler, and for science, a few other key individuals were more discerning and less supercilious than Dr Bayer. One was Herman Mark, the polymer pioneer we met earlier (Ch. 2). By this time, Mark was an established leader in the world of polymers and had a busy schedule of lecturing and consulting. 'My role in this whole affair', he told me, 'was that of a post-box'.

But 'prophet' and 'publicist' might be more appropriate terms ('Geheimrat', i.e. 'Privy Councillor', is applied affectionately by his colleagues at Brooklyn Polytechnic Institute, which Mark put on the world polymer map). He constituted a one-man Early Warning System

---

*Ziegler, Holzkamp, Breil and Martin, *Angew. Chem.* **67** No. 19/20, 541–636 (1955).

on new developments in polymers for a number of research laboratories for whom he consulted. ('But at duPont', he used to say, 'my chief function was to tell one department what the other duPont departments were doing.')

A personal friend of Karl Ziegler's since 1927, Mark always stopped at Mülheim on his semi-annual trips to Europe to learn what was going on in its laboratories. On one of these occasions, Ziegler surprised Mark with a request. 'Tell me something about polymers', he said.

He had already discovered the 'Aufbau' reaction and was already wondering how far it might lead.

Mark thus heard the whole story first-hand. Unlike most first-time hearers, he was not turned off by the low and variable molecular weight of the early products. His intimate familiarity with polymers enabled him to appreciate the significance of the completely unbranched structures that were formed, and he was characteristically optimistic that ways might be found greatly to increase chain length. 'New polymers always arrive as babies and have to grow', he explains.

Mark was sufficiently impressed that when he arrived in England he made it a point to describe Ziegler's work to another eminent friend, Sir Robert Robinson. Sir Robert had become a towering figure in the history of organic chemistry in England and a Nobel laureate. He had also become a consultant and board member for Petrochemicals Ltd, a young British company trying to build up a business based on chemical derivatives of olefin hydrocarbons derived from oil and gas.

In the best British tradition, their discussion was held over a glass of sherry in Sir Robert's London club, the 'Athenaeum'. It was a historically significant meeting, for Sir Robert, who had had a hand, twenty years earlier, in planning the experiments that led to the accidental discovery of high-pressure polyethylene (Ch. 2), thereby also became an active participant in the events that later converted Ziegler's accidental discovery of linear polyethylene into one of the most fruitful innovations the world has ever seen.

Mark recalls that Sir Robert immediately agreed with his colourful prediction that 'a big snake may grow from this little worm' (i.e. the chain length may be greatly increased). Robinson denied to me that he foresaw the ultimate results, but he recognised Ziegler's process as something new one could do with olefins and therefore something potentially important to PCL's business. He therefore persuaded the other members of their board of directors to let him dispatch Dr E. T. Borrows, their research director, to Mülheim to get the full story and, if possible, rights for research and commercial development in England.

Borrows, an outgoing person who is given to direct action and has well founded confidence in his own persuasive powers, promptly telephoned Mülheim to get an appointment with Prof. Ziegler. Instead of the Professor, he got Frau Professor, who told him that Herr Dr Director Ziegler was off on a mountain climbing holiday in Switzerland with his son and his chief assistant, Dr Wilke. Most people would have accepted delay as inevitable, but not Borrows. He fired off cables that eventually won agreement that he could meet Ziegler in Sils Maria, a lovely alpine village near St Moritz.

Taking along a German-speaking assistant to help him over the language hurdles, Borrows arrived early in Sils Maria and spent the day waiting for Ziegler to come off the mountain. In the evening, they met and began their discussions over dinner. Borrows sensed that Ziegler was interested in licensing even at this early stage because the precedent would strengthen his hand in later negotiations with other people for other parts of the world*. But he also quickly perceived that Ziegler's natural inclination was to rely less on binding legalisms than on a soundly built-up basis of mutual respect and trust between individuals.

So Borrows undertook his own 'build-up reaction', remaining persistent without being pushy. On the second day, he joined Ziegler in going out to meet Wilke and young Ziegler as they came down from climbing over an intervening mountain and spent a pleasant day enjoying the company and the scenery, with only occasional talk of licences. Ziegler played the game a little with Borrows, hinting that another company (probably the Italian chemical company, Montecatini, of whom more later) was also interested. But by the time he returned to England, Borrows had agreement in principle on licensing terms, and PCL's board was soon persuaded to pursue this opportunity. In subsequent trips to Mülheim, wordings were hammered out, first of a letter of intent and finally of a firm licensing agreement that gave PCL exclusive manufacturing rights for the U.K.

Ziegler's distrust of lawyers was equalled by his confidence in his

---

*As we shall see from later events, this proved sound strategy. The sheep-herd syndrome is more prevalent in industry than its leaders would acknowledge, and there is no better way to stimulate a prospective licensee's interest in a new process than to have him discover that a competitor is after the same prize.

own abilities, so he undertook to conduct all negotiations personally. He told Borrows that he preferred to deal on the basis of mutual trust and understanding between themselves, but that if Borrows showed up with a couple of lawyers in tow, he, Ziegler, would have to get a couple also.

Ziegler's espousal of mutual trust and gentlemen's agreements was undoubtedly sincere, but he also probably felt that he had the best chance of getting what he wanted in a man-to-man situation. He had already held licensing discussions with several firms inside, and at least one outside, Germany, and this doubtless added to his confidence. Certainly, he did not let mutual trust substitute for spelling out what he wanted in exact terms. The story has been told that he took home law books from the Mülheim public library to help him in drafting both licences and patent applications (more successfully for the former than for the latter, as we shall see). He argued strongly with Borrows over the language to be used in the agreement (German *v* English) and the meaning of specific phrases, and insisted on a requirement that the licensee could not 'sit on' his licence but must immediately undertake large-scale development. This last provision was to cost PCL and, later, Shell millions of pounds and uncountable headaches, but it achieved its purpose in getting a product (of sorts) on the market at an early stage.

The final document created by these two non-professional negotiators (although, in Borrows' case, with professional assistance and management resistance back in England) is a noteworthy one. Not only was it the work of amateurs; it was the first of Ziegler's many licences outside Germany and one of only three exclusives he ever granted. It stood up to the stresses laid upon it later, and its phrases were broad enough (at Borrows' insistence) that they stood PCL in good stead when unforeseen discoveries at Mülheim greatly broadened the field and escalated its significance.

In the U.S., the minds most receptive to Ziegler's story of the 'growth reaction' were found, somewhat improbably, in the Hercules Powder Company. That company was one of the splinters from the break-up of the original duPont Company, which in turn had close ties with ICI in England. Although separated from such direct sources through anti-trust actions, Hercules' management retained a keen awareness of the pre-eminence of European, and particularly German, chemistry and its relevance to the company's main goal. That was to diversify from explosives into industrial chemicals. Like duPont, Hercules had long since read and heeded the handwriting on the wall for explosives manufacturers and had turned their hand to applying the same technology to the manufacture of plastics by chemical modification of cellulose (cellulose acetate, butyrate, etc.)*.

The company had a good 'European Early Warning System', with an office in The Hague and frequent visits by executives from the head office in Wilmington, Delaware. By 1950, some of those executives, and particularly David Wiggam, recognised the need for a broader base in the plastics business and were looking for opportunities to move into newer, all-synthetic polymers.

Wiggam was travelling in Europe in 1950 with Dr Arthur Glasebrook† from Wilmington and their representative in The Hague, Max Riemersma. The trio made a business call on a machinery firm in Mülheim and, while there, Riemersma suggested a 'courtesy call' on Dr Koch, a senior scientist at the Max Planck Institute. When he learned of Hercules' potential interest in new polymers, Koch remarked that Prof. Ziegler was doing research on polymerisation of ethylene.

The Hercules representatives did not immediately follow this fortuitous lead. It was not until a year later that Wiggam met Ziegler, and it was then in connection with an entirely different reaction (use of Ziegler's catalysts to convert ethylene to *p*-xylene, a starting material for making dimethyl terephthalate, a precursor for polyester resins and fibres). However, Wiggam took a broad approach, negotiating for an agreement that would give Hercules an option on exclusive rights in the U.S. to all of Ziegler's chemistry based on metal alkyls. General agreement was reached eventually on terms and conditions, but Ziegler showed reluctance to put his name to paper.

At this critical juncture, in the summer of 1952, Wiggam was temporarily confined to a hospital, but he took advantage of the presence in Europe of Paul Johnstone, an aggressive and articulate executive who was then in Hercules' Synthetics Department but was destined to

---

*Reflecting this change in business orientation, the corporate name eventually became simply Hercules Inc.

†Like Calvin Schildknecht (Ch.2), Glasebrook had obtained his Ph.D. at Johns Hopkins while Maurice Huggins was teaching there. He came to the Hercules Research Center in 1942 and headed several different research divisions before becoming an unofficial 'overseas representative' in 1953 (his formal title at that time was Assistant to the Director of Research, but it was changed (belatedly?) in 1961 to Overseas Technical Manager).

become a vice president and play a key role in subsequent licensing and commercial development of polyolefins for Hercules and U.S. industry. Johnstone visited Mülheim to try to help bring the negotiations to completion. In talking with Ziegler, he discovered that the latter was looking for a way to finance a lecture tour and visit to the United States for himself and his wife. When Johnstone astutely suggested including underwriting of such a trip in the terms of the licence agreement, this 'bonus' provided just enough added inducement to persuade Ziegler to sign the option.

This success gave Hercules an opportunity similar to those enjoyed by Petrochemicals Ltd in England, Hoechst in Germany and Montecatini in Italy, but surpassing them in ultimate potential. What they did with that option and opportunity will come into our story later.

4 Giulio Natta and the Milan Polytechnic Institute

Natta, the man

Giulio Natta is a complex personality and brilliant mind, housed in and shielded by a disarmingly mild exterior. A quietly dressed, soft-spoken man of short stature and gentle mien, he has consistently displayed courtesy and consideration toward others, but a deep intensity constantly showed in his eyes. On first impression, one tended to categorise him as the traditional dedicated pure scientist. While this is not untruth, it is so far from complete truth that a professional colleague who knew him well, on hearing me give such a description, smiled and said: 'Oh, so you have been fooled, too!'

Indeed, the record shows that Natta's professional activities and personal interests encompass the practical fully as much as the theoretical, the traditional as much as the modern, and ambition for self as much as for science. Only a complicated man could achieve the outwardly smooth blending of such diverse elements.

To understand a nation or a man requires knowing something of origins. In 1903, in Imperia, in the region of Ligure, Giulio Natta was born into the heart of one of the oldest cultural communities in Italy, and into a family that is part of that culture and its traditions. The 'Liguri', by ancient tradition, were part of the neolithic culture that first fashioned pottery and weapons in northern Italy ten thousand years before Christ. Nine millenia later, their successors, the Ligures, were among the independent tribes who, with the Etruscans, battled the armies of Rome to delay, and the Celtic hordes to unwittingly aid, the birth of the Roman Empire.

Ligure is within a hundred kilometres of Turin and of Genoa with their special cultural and historical traditions, but equally close in the opposite direction are Milan, hub of industry and technology, and the oil fields of northern Italy, part of the economic base that underwrote the technology that made Natta's career possible.

In modern times, the traditional, almost the only, career in the Natta family was the profession of law. Giulio's father, his uncles and his cousins were all magistrates at one time or another. The son was expected to follow this time-worn and time-honoured path, and he would probably have done so had not another family tradition unintentionally turned the young man into a byway that became his highroad to renown.

As part of his early training, each (each male?) youth of the Natta family was assigned the task of exploring and reporting on some particular field of human knowledge. Young Giulio's assignment happened to be chemistry. And so it was that at the age of twelve he 'bumped into chemistry' (his words). The new world thus encountered was so fascinating that he was 'bumped' out of his predestined path and into the pursuit of science as his life work. As might be guessed, this defection was a disappointment to his parents, who referred to him thereafter as the 'white fly' of the family*.

But the road that ultimately led Natta into polymer science was as multibranched as were polymer chains prior to his and Ziegler's researches. After attending 'Christopher Columbus' high school in Genoa, he began his scientific training with a curriculum in pure mathematics at the University of Genoa. He changed disciplines as well as schools when he moved to Milan Polytechnic, where he took his 'Dottore' degree in Chemical Engineering at the nearly unprecedented age of twenty-one. This achievement was followed by the award of the 'Libero Docente' degree of the same institute three years later.

Journalistic accounts credit Natta with having recognised, while still a teenager, that while chemistry was his love, chemical engineering would be a better helpmate, i.e. that a Ch.E. degree would be 'indispensable' to a successful career. Such a keen and pragmatic insight scarcely

---

*This metaphor (*musca bianca*) is presumably analogous in its implication of non-conformity, but without derogatory connotation, to 'black sheep'.

fits the image of a young, theoretical, 'pure' scientist. Years later, when asked about the secrets of his success, he said: 'First of all, I had the great luck to have a degree in Chemical Engineering'.

That he chose to make such an answer is a good indication that it was not luck at all.

On the same occasion, Natta said that he entered the University only because it offered him a laboratory, and that he never intended to pursue an academic career. Instead, as he said:

'I always sought to work alongside of industry. Industry raises practical problems. The secret is to solve them with scientific methods. If that takes place, the result is always positive. Industry taught me always to seek the practical side in any research'.

The 'positive results' were certainly forthcoming in his case and included peer recognition, financial rewards and support for further research, all highly gratifying to this practical and ambitious scientist.

Ability, teamed with determination, advanced Natta's career in impressive fashion. Starting as an Assistant Professor at Milan, he moved to the University of Pavia as full Professor and Director of the Institute of General Chemistry; then to the University of Rome as Professor and Director of the Institute of Physical Chemistry. The year 1937 found him in Turin as head of the Institute of Industrial Chemistry. Finally, he returned to Milan in 1938 as Professor and Director of the Milan Institute of Industrial Chemistry.

Interestingly, Natta has said that the 'autocracy' (euphemism for Mussolini's dictatorship) gave indirect assistance to his ambitions for a research career. This result came from the heavy-handed support given by the fascist regime to 'research' because of the useful technology it was expected to spawn, and from the aftermath of that policy. Technically trained people were pushed indiscriminately into research, many against their natural abilities and inclination; consequently, when the 'autocracy' left the political scene, the research draftees left the scientific scene, abandoning the field and the improved facilities to the truly committed.

Throughout his early career, Natta's varied professorial duties were accompanied by varied research experience, which he credits with preparing him for his subsequent remarkable contributions to polymer science. His early (1924) research familiarised him with the use of x-rays and electron beams as elegant and powerful tools for elucidating the secrets of the structure of solids, and he applied them successfully to studies of solid catalysts and certain polymers.

His work with electron diffraction in 1934 marked the first use of this sophisticated technique in original research in Italy. He also studied the mechanism (kinetics) of such reactions as the synthesis of methyl alcohol from hydrogen and carbon monoxide, synthesis of higher alcohols from the same gases and olefins (oxo-reaction), and the addition of hydrogen to unsaturated compounds.

It was significant for Natta's subsequent career, and of historical interest to us, that he was introduced to the world of high polymers by Hermann Staudinger, the man who, more than any other, was the founder of modern polymer theory (Chs 2, 3). This encounter, which occurred in Freiburg, Germany, in 1932, aroused Natta's interest in problems of polymer structure and provides further evidence that Staudinger's influence is as the shadow of a giant in the history of giant molecules.

The first significant work in Natta's own laboratory, related directly to production of high polymers, was in the field of synthetic rubber; it dealt with the purification of butadiene, the most important starting material. (The process was the then-new extractive distillation method for physical separation of butadiene from other hydrocarbons.) At the same time (1938) he began studies of the 'polymerisation' of olefins (reactions that today would be called oligomerisation: combining just a few molecules to form very low polymers, mostly dimers and trimers).

The career summarised above reads like a fairly orthodox story of success in science, unusual only in the rate and reach of achievement it represents. Since science is one of the jealous professions, the price for a record like Natta's might well have been a single-minded, full-time devotion to modern chemistry, to the exclusion of almost everything else. But Giulio Natta had numerous other personal interests all his life, interests that stand in such pointed contrast to the disciplines of 'hard science' as to suggest that they were a conscious or subconscious counterpoint to, if not a partial negation of, his commitment to chemistry.

There is a saying in Italy that 'all Ligurians are nature lovers'. This is one tradition that Natta honoured in full measure. While still a student, he became an enthusiastic 'caveman' (*spelunker*). He was also an active mountain climber as a member of the local alpine club, and continued mountaineering until his professional activities became too demanding. Even then, he went to the mountains for recreation–hunting, fishing and ski-ing. His mountain lodge near St Moritz was a 'retreat' in a real sense, and his scientific colleagues were not normally included in these non-science breaks.

Another hobby was collecting fossils, and evidence that he was successful at it was displayed in his home in Milan. His study was also decorated with eighteenth century paintings and a large portrait of his father, the judge, in his ermine-trimmed magistrate's robes. But the hobby that Natta pursued longest was collecting mushrooms. Long walks in the woods were accompanied by the collection, and followed by the consumption, of a great variety of edible fungi. It fell to Mrs Natta to prepare these delicacies for the table and to join in their enjoyment, even though they often involved new and strange varieties. She insisted that she had no qualms, fully trusting her husband's judgment as well as his taste, and neither ever proved wrong*.

Despite the intensity he displayed when involved with his favourite subjects, Natta maintained a quiet self-control that precluded emotional displays. But the logical, disciplined scientist was not without sentiment or sensitivities. For instance, he did not drive, nor would he willingly ride in an automobile, a trait attributed by some who knew him well to the fact that his mother was killed in a tragic automobile accident. Ironically, the only practical way of travelling on his frequent trips to his father's home at Sanremo (on the eastern end of the Italian Riviera) was by automobile, and the route led through the town where his mother was buried. He never failed to stop there to visit her grave.

Natta and Ziegler: career parallels

It is striking to note the parallelism in the careers and interests of the two men who came so close together and fell so far apart while they were making history in stereoregular polymerisation. Born within a few years of each other, both came from families with a tradition of classical, non-scientific professions (the clergy and the law). Both were lured away from those traditions and into science by reading about it during boyhood: Ziegler at age eleven, Natta at twelve. Each showed precocious ability as a student and completed his academic training at a phenomenally early age (Ziegler–doctorate at twenty-one, 'Privatdocent' at twenty-three; Natta–'Dottore' at twenty-one, 'Libero Docente' at

---

*Rosita Natta was something of an outdoor person herself. At a research conference in the British Isles, she spent the entire period fishing in the local streams.

twenty-four). Each started teaching at his *alma mater* but soon went to another school. While Ziegler's career was taking him through three other German universities, Natta was progressing similarly through three Italian universities. Each ended up heading his own research institute, financed in part by contributions from industry.

Although their personalities were quite different, some of their personal interests also ran parallel: most noticeably, the strong love for mountaineering that both exhibited through most of their active lives. Incidentally, this passion was shared by at least one of the other key figures in this history, Sir Robert Robinson. He became a good friend of both Ziegler and Natta and he himself was still going into the mountains in his seventies. It would add a dramatic touch to be able to report that all three of these Nobel Laureates, or even any two, had shared a mountaintop experience at some time, but there is no record of this having happened.

Some of the parallels cited above, however intriguing, are simply coincidence. But others are significant because they seem typical of individuals destined to be great achievers in science. The very early commitment to science as a life interest, the outstanding scholastic record, the rapid rise within the chosen profession, all are milestones that mark the path of many (although by no means all) science greats. It can even be argued that the love of mountaineering is representative of a tendency for research scientists to favour this type of recreation over more organised team sports.

> The argument runs to the effect that team sports (football, soccer, baseball, hockey) have more natural appeal to businessmen and others involved in group efforts because, like business, they are played out within a designated area according to prescribed and arbitrary rules; and superior achievement depends, not on great originality and innovative approaches (innovation being essentially proscribed by the rules), but on execution of standard repetitive operations with superior skill*. Furthermore, success (winning) means defeating one's opponents (competitors). In mountaineering and in research, on the other hand, it is not the defeat of human opponents (save in the sense of getting there before others do), but

---

*Businessmen and some coaches will protest that being highly innovative is the key to their success, but an objective view of what is acclaimed as innovation in such cases will, with rare and notable exceptions, show it to be a vernier adjustment within the rules of the game.

the challenge of nature's unknown and unconquered heights that provides spur to supreme achievement. Teamwork is often involved, but at the apex it is the ability of the individual to solve the problem posed by nature that is the measure of his accomplishment*.

If this hypothesis has merit, it should be possible, by proper sampling and analysis, to demonstrate that outstanding scientists are proportionately more likely to be found on a mountain than are persons of contrasting occupations. The sample at hand is, of course, far too small to be significant; the set of all Nobel Prize winners in science might make a manageable and valid starting point for a sociological thesis.

But there is no need for speculation regarding one outstanding attribute shared by both men and clearly recognisable as a vital key to their achievement of great prominence. Although steeped in, and dedicated to, pursuit of science, each had a strong and unabashed appreciation of the potential for translating scientific discoveries into practical benefits for industry and mankind—and themselves.

Natta and Giustiniani

Since earliest times, it has been traditional in Italy for genius to have a 'patrono'. DaVinci had his Duke, and Michelangelo his Pope; Giulio Natta had Dr Ing. Piero Giustiniani, Managing Director of the Montecatini Chemical Company. This large, shrewd, vainglorious man worked his way to the top of Italy's largest chemical company (and eventually out of it) by submerging everything else to his ambition and by being a business busybody—trying to know everything about everything and to decide everything himself. Technically trained himself, he recognised in Natta a rare blend of modernist with traditionalist, theoretical scientist with industrial-minded engineer, that could be of great value to his, Giustiniani's, company.

He therefore arranged for Montecatini to retain Natta as a consultant and to support financially the Institute of Industrial Chemistry at Milan Polytechnic, with Natta as its head. Some promising young technical men hired by Montecatini were also sent to Natta for further

---

*A similar analogy has been drawn, with humorous effect, between the nature of the research scientist and his preference for a 'loner' musical instrument (the 'cello) rather than a 'group' instrument (piano) in the play 'Absence of A 'Cello' by Ira Wallach. As with most real humour, it is underlaid with truth. In our case, the protagonists all enjoyed classical music, but since they are not noted as performers, this test cannot readily be applied.

training. In return for this support, Natta assigned to Montecatini world-wide rights to patents and commercial exploitation of any discoveries in fields of interest to that company. This was no superficial or one-shot interest on Giustiniani's part; the two men became very well acquainted, and Giustiniani took a personal interest in the possibilities for commercial development of the findings of his consultant and friend.

How deep the friendship went is difficult to say today, for even close observers differ in their assessments. The two men were so different in personality, 'style' and even size, that it is hard to believe that it was not, at least on Natta's part, *une amité de convenance*. But we cannot be sure; 'Mutt and Jeff' relationships are not confined to comic strips. What is certain is that the relationship was close and the influence strong.

There was thus a striking contrast between Natta's tight contract and close working relationship with Montecatini and Ziegler's 'blue sky' contract and arm's length dealings with his sponsors. To recognise that contrast is more than interesting, it is crucial to understanding later events. No team, horses or men, can pull smoothly together when the reins are taut on one side and loose on the other.

A plaque mounted on the wall of 'Natta's Institute' records its indebtedness to Montecatini for support and symbolises the strong Montecatini 'presence' that accompanied it. Inevitably, there were restrictions on publication; however, these were exercised in a way that still permitted a record of early and extensive publication, generally high in quality.

Natta encounters Ziegler

In the years before their interests converged, Ziegler was working on metal-organic compounds and Natta on reactions of hydrocarbons, consistent with Montecatini's interest in petrochemicals (commercially important chemicals derivable from then-cheap and abundant refinery gases). What brought them together was Ziegler's discovery of the 'Aufbau' reaction (Ch.3) and Natta's work on 'polymerisation' of olefins to dimers and trimers. In 1952, Natta learned of a lecture Ziegler had scheduled in Frankfurt. Although that lecture did not create any general stir, Natta attended, and was stirred.

Ziegler subsequently gave similar lectures at a number of places around the world and before many skilled scientists, but, with a few notable exceptions, as mentioned above, scarcely any appreciated the significance of what he was reporting. Natta was among the few who did, and he acted promptly. Afterwards, he said of that meeting:

'The knowledge acquired in the field of the polymerisation of olefins enabled me to appreciate the singularity of the methods . . . that Karl Ziegler described', and, 'my interest was aroused.'

Natta also managed to arouse Montecatini's interest. The Institute budget was so tight that Natta could not offer Ziegler a rail ticket, let alone an honorarium, so he persuaded Giustiniani to finance an invitation to Ziegler to visit Milan for a discussion of science and business. Out of that meeting came an agreement whereby Montecatini purchased rights for commercial development of Ziegler's research in Italy, and Natta acquired access to Ziegler's continuing research, in a field described as the 'transformation of olefins'. Ziegler could thus take considerable satisfaction in having won recognition by scientific peers plus significant financial compensation from a major chemical company for developments that were still in the laboratory stage. And under the terms of his contract with the German coal companies, the money would be his!

In the happy climate of a cooperative research agreement which both recognised as mutually beneficial, professional respect between the two scientists soon ripened into friendship. In time, their meetings became social as well as business occasions and, as their wives were included, Mrs Ziegler and Mrs Natta also became good friends. The two families even spent some vacation time together.

The 'Three Bright Boys'

As part of the cooperative research and licensing agreement, it was arranged that research people from Milan would spend some time in Ziegler's laboratory to 'learn the process forming an object of the newly signed contract'. In February 1953, the following three men were despatched to Mülheim: Giovanni Crespi, chemical engineer; Paolo Chini, chemist; and Roberto Magri, chemist.

Although these were Natta's research representatives, they were Montecatini's men. They had been hired by that company only a year or two previously and had been assigned to work at Natta's Institute as part of their training for an eventual career in Montecatini. While at the Institute, their work assignments and direction were wholly under Natta's control; but they were subject to recall at Montecatini's discretion and they knew that their futures, if they met expectations, lay with Montecatini. While visiting Ziegler's laboratory, their expenses were approved and paid by Montecatini, not Natta.

Crespi had obtained his degree at the Milan Polytechnic Institute, and Magri had spent a year there on a scholarship. He and Chini had obtained degrees at the University of Florence and had become friends there.

Once in Ziegler's laboratory, these young men were given appropriate assignments within the field of the agreement. They were free to mingle with the research staff, but they were not privy to the entire Institute program, and were expected to confine their enquiries and activities to the agreed-on area. This was broadly defined in the contract as the 'transformation of olefins' but consisted *de facto* in making dimers and low polymers (oils and waxes) from the simple, lower olefins. Nevertheless, as might be expected of curious and ambitious young men on their first 'outside' assignment, they absorbed whatever information came their way through either official or unofficial channels and sent private reports back to Milan.

Thus it was that Natta and Montecatini had representatives on the scene and a 'hot line' for information from Ziegler's laboratory at the time the momentous discovery was made that stretched the scope of the original program and severely strained its working mechanism.

5 Fruitful Innovation — 1. The Polyethylene Discovery

> It was an event of great importance. . . . There has never been any chemical process whose disclosure has aroused such interest in all the history of technology.—Topchiev and Kreutsel*

The 'nickel effect'

Discovery of the 'growth reaction' in Ziegler's laboratory opened the way for the eventual discovery of linear polyethylene, since the latter represents only an extremely long continuation of the stepwise chain growth process of adding one molecule of ethylene after another. To paraphrase the old Chinese saying: 'The longest polymer chain begins with a single step.' But the extrapolation was extreme, analogous to extending a trip to the corner drug store to a cross-country journey, or an Aerobee rocket flight to the Apollo 14 mission. Despite its remoteness, however, the possibility of making a high polymer was not entirely unrecognised as at least a distant goal.

Dr Heinz Martin, one of the more senior staff members at the time, recalls that, as soon as the possibility of running the growth reaction as a catalytic process (using huge excess of ethylene) was demonstrated, the immediate question was asked, 'How far can we go?' and, at least by implication, 'Can we produce polyethylene?' Thus, although the

---

*Topchiev and Kreutsel, *Polyolefins,* Pergamon, New York (1962)

discovery of the successful process was accidental, the fact that such a process was possible did not come as a complete surprise.

As so often happens in research, the key initial experiment was made in pursuit of a quite different objective, and the critical factors were an accidental contaminant and an alert observer.

E. Holzkamp, at that time (1952–53) a doctoral candidate at the Institute in Mülheim, had been assigned the task of finding out how to make odd-numbered olefins in the growth reaction.

Starting with ethylene and aluminium triethyl, only olefin chains of an even number of carbon atoms could be expected (C_2 + C_2 + C_2 etc.). With a one- or three-carbon aluminium compound, however, adding ethylene might give odd-numbered chains if the alkyl group participated (C_3 + C_2 etc.). The three-carbon aluminium alkyl, aluminium tri-isopropyl, was the easiest to work with, so Holzkamp used it. Reacting it with ethylene in a steel pressure vessel, he was startled to find that he got neither the desired odd-numbered nor the usual mixture of even-numbered olefins, but almost exclusively the four-carbon olefin, 1-butene. This meant that the growth reaction had not grown past the first step (the dimer stage); every four-carbon group had been knocked off the aluminium atom the instant it was formed. Evidently, the displacement reaction had somehow been tremendously accelerated.

Holzkamp went back to the normal system of aluminium triethyl and ethylene for a check, but when he ran it in the same vessel he again got butene.

The inexplicable result encountering a fertile mind is often the father of new understanding, and so it proved in this instance. Systematic consideration and successive elmination of all other possible variables soon showed that the particular metal vessel used in these experiments was responsible for the unique results. Further sleuthing showed that its special influence on the reaction derived from the fact that it carried traces of dissolved, or colloidally dispersed, nickel metal.

At this point, published and private accounts of events and causes are surprisingly divergent. I was once told that the vessel had a nickel lining; another and more widespread story is that the reactor had previously been used for a reaction (hydrogenation) in which very fine nickel powder was added as a catalyst. It was even said that it was another chemist in the laboratory who called Holzkamp's attention to the reactor's previous history and suggested nickel as the active contaminant.

But the most convincing, because the best-documented, account has it that, while suspecting residual hydrogenation catalyst may have been useful in turning attention to nickel, that did not prove to be the actual source of the metal. Rather, it was finally determined, it came from the nickel-bearing stainless-steel walls of the reaction vessel itself. An elaborate (seven-step) and aggressive cleaning procedure was employed routinely between experiments. A tiny amount of nickel was dissolved from the stainless steel by a nitric acid wash (step 2), was converted to an insoluble residue in a later alkaline wash (step 4), and was finally re-precipitated as finely divided nickel by the reducing action of aluminium alkyl added in the next experiment.

Only minute traces of active nickel could have been produced and retained, but that was sufficient. The clincher, naturally, was the finding that deliberate addition of nickel salts produced the same effect, even in a glass reactor.

This abbreviated description of the famous 'nickel effect' discovery is at least consistent with the published statements of Ziegler himself, with the account given in Breil's thesis, and with the personal recollections of Heinz Martin. But the most direct support comes from Arthur Glasebrook, Hercules' 'observer in residence' at Mülheim that summer, along with the representatives from Montecatini and other prospective licensees. Having heard of the discovery from Wilke, Glasebrook had a talk with Halzkamp himself and then (June 1953) fired off a memo to his home office giving an explicit description of the tests and results that is better documentation than any of the Institute's own records.

The in-house records of that particular period are unusually skimpy for an otherwise so meticulous group. Things were moving so quickly just then that there did not seem to be time to get everything written down before the next experiment had again changed the picture. Ziegler's group was never famous for detailed, 'on-time' reporting, and one senses the pressure from the 'Herr Director' who, having discussed the results with his student, was much more interested in having him test the next idea than in reading about the last one.

The seemingly minor details which we have dwelt on here are of interest because of the remarkable string of coincidences that was necessary for the discovery to be made at all: use of a nickel-bearing stainless-steel reactor, a particular multistep cleaning procedure with the steps in a particular order, a reaction ingredient that reduces nickel salts, and a reaction extraordinarily sensitive to traces of nickel. It is also an example of good scientific detective work and a classic negative example of the importance of diligent record keeping, especially in those periods of active discovery when it is more difficult. The true importance of the experiments and the records became evident only later on, after the 'nickel effect' had led to a far more momentous discovery.

The 'nickel effect' by itself was a minor milestone, but it marked progress in the opposite direction so far as getting to high polymers was concerned. Rather than accelerating the displacement reaction, what was needed was a way to suppress it. Better yet, the growth reaction needed to be speeded up*, so that the molecular chains would grow both quickly and to great length before being 'killed' by displacement from the aluminium atom.

Actually, at that moment the Institute approach probably was simply to follow up the lead that had been opened. One metal, nickel, had been found to be a co-catalyst that profoundly affected the growth reaction; what other metals might have what similar or different effects? In Ziegler's words, the obvious thing to do was a systematic work-through (*systematische durchmustern*) of the periodic chart of the elements, testing every metallic element in turn.

The first high polymer

This course of action, to quote Ziegler again, soon proved the truth of the adage '*les extrèmes se touche*'. In this instance, the extremes were connected in the sense that learning how to make the lowest of all ethylene polymers, butene, was the turning point which quickly led to learning how to make the highest polymer, the 'true', linear, polyethylene.

Holzkamp started the systematic search, naturally, by looking first at the metals closely related to nickel on the chemical family tree. He placed orders for the available metal compounds from chemical suppliers, but meanwhile he looked on the Institute's storeroom shelves to see what he could lay his hands on immediately. He found a chromium

---

*The possibility was recognised that traces of some co-catalyst had been present, undetected, in all previous experiments, and that scrupulous purification might eliminate or mitigate the displacement reaction. This idea was tested but did not work; thermal displacement still occurred and limited the chain length.

compound, so he tried it forthwith*. In the first test, it gave a mixture of butene and some higher olefins–not a very exciting result. Fortunately, however, a second run was made, and this gave a more interesting mixture of products: some butene and other olefins, but also a small amount of high-molecular-weight (long-chain) material–a 'real' high polymer.

This was an intriguing result, despite the mixed product, but its full significance was not immediately appreciated. In fact, the person who took greatest immediate note was probably again the outside observer, Arthur Glasebrook. He had visited Holzkamp's laboratory for further discussion of the 'nickel effect' and learned first-hand of the results Holzkamp had just obtained with chromium. He promptly put the essential facts on a wire recording which he mailed to Hercules' home office. However, even though his report included the statement, 'the product may be very interesting', there was no immediate follow-up from Wilmington. Thus another opportunity to act on early information was missed. But Glasebrook's report later proved useful to Ziegler in establishing dates in a patent interference action.

Although Ziegler recognised that a new lead had been opened up, he also recognised his responsibilities as Herr Professor. He knew that, for a young man preparing for the examinations and doctorate thesis that would make or break his future as a scientist (German doctorate examinations being justly famous as real ordeals), completing that work was of pre-eminent importance. So he did not redirect Holzkamp's own research in this new direction; instead, he assigned it to a still younger doctoral candidate, Heinz Breil, and told Holzkamp that Breil would be under his general supervision†.

---

*Iron compounds were also doubtless at hand, but iron was known to have little or no activity, since iron and glass vessels gave similar results. Subsequent tests showed that iron actually slightly suppressed the effect of nickel.

†This was something of a departure from Ziegler's normally strictly 'horizontal' organisation, although not unprecedented. In any event, the degree of supervision actually exercised by Holzkamp must have been modest. In Breil's thesis, although Holzkamp's prior work is fully described, the formal acknowledgement page carries only the traditional expression of appreciation to 'my highly revered teacher, Herr Prof. Dr Phil. rer. nat. e.h. Karl Ziegler'. Breil himself was the recipient of the same treatment later when he was directed by Ziegler to complete the theoretical studies that formed the theme of his dissertation, leaving to others the pursuit of the exciting technological questions.

But Herr Dr Ziegler, the scientist and laboratory director, was also true to the German tradition of a logical and thorough approach to problem solving. He soon reorganised the greater part of the Institute force* to bring nearly all its brainpower to bear on a systematic development and unravelling of the new phenomenon. Heinz Martin undertook the systematic variation of the reaction conditions, along with studies of the reaction mechanism, in order to improve the process and drive the chain length of the ethylene polymers as high as possible. Other chemists were assigned to further catalyst research, still others to the so-called pilot plant, etc.

I have heard it suggested that such a systematic approach prevented any 'intuitive leap' that might have gained much new ground quickly. Conversely, the criticism has also been voiced that the elegant and powerful tools of modern physical chemistry were not applied with the skill and force that could have illuminated the kinetics, and hence the course, of the reaction, the nature of the catalyst and the structure of the product.

Neither was chemical engineering much in evidence, i.e. quantitative studies of reaction steps and engineering problems that would permit rapid scaling up and plant design. No apparatus was available for studying reactions under continuous, steady-state conditions; however, uncertain monomer purity would have made rate measurements of dubious value in any case. Process engineering was virtually non-existent.

The Institute's 'pilot plant' was primarily a facility for scaled-up preparations, and was used mostly for working out and applying methods for preparing aluminium alkyls on a larger scale than was convenient and safe in the laboratory. Safety was a real problem; Ziegler understated the case when he wrote: 'the properties of these catalysts are not exactly commonplace, and handling the concentrated pure alkyls calls for a little practice.' Among many non-commonplace properties: they react explosively with water and burst into flame spontaneously on exposure to air.

Retrospective criticism, always easy, is always suspect. Critical connotations aside, however, the comments quoted above probably have

---

*The major exception was the 'Fischer Section', comprising groups working on certain projects Ziegler had inherited from his predecessor, the renowned Prof. Franz Fischer. Because of the well publicised important contributions that had come from these groups in the past, Ziegler found it politic to let their work continue undisturbed.

considerable descriptive validity. Germany was the father of organic chemistry, but physical chemistry was a second-generation stepchild, and chemical engineering was a recent immigrant whose parentage was clear but whose legitimacy remained to be established. Despite his originality and his own interest in physical chemistry, Ziegler himself was fundamentally an organic chemist of the classical German school, and the Institute was an extension of himself. His approach to research problems was inductive, rather than deductive. When asked after a lecture what he thought certain types of molecules ought to do under given conditions, he replied: 'I don't know what they *ought* to do; I only know what they *do* do.'

On another occasion, when he was asked if he had attended a lecture by Prof. Natta, he replied: 'No, I never attended such a meeting. I preferred to make new discoveries in polymer chemistry more than to listen to lectures in polymer chemistry.' (This, from the man who, just a few years earlier, had asked Prof. Mark to 'tell him something' about polymers!)

Ziegler's overriding interest was in the chemical reaction and its products. A friend and contemporary once complained that 'Ziegler never had much interest in mechanism; he never wrote an equation.' This is an exaggeration; as noted above, kinetic studies of a sort were included in Heinz Martin's program at an early stage.

The first feasible process

In any event, it is hard to be critical about the organisation and approach in face of the fact that they were undeniably successful and highly productive. For a time, in fact, progress was too rapid to be orderly. Within a few months of the first experiment that made any high polymer at all, catalysts and conditions had been found that gave quantitative yields of polymer under unprecedentedly mild conditions, the most active form (valence state) of the metal co-catalyst had been established, and control had been achieved over the molecular weight of the polymer.

This burst began soon after Breil started on what looked like an unimaginative plodding through the list of available metal compounds. Breil, in fact, confided to a friend that he was at first quite disappointed in the assignment given him by Ziegler. But to a researcher, happiness is a new phenomenon, so Breil's unhappiness was short-lived, dissipating when the first exciting results appeared. According to Ziegler's own description, Breil's work, after an unexciting start 'suddenly, within a few dramatic days, almost within hours, led to the discovery of the new polyethylene process'.

A number of metals (cobalt, copper, iron, silver, gold, platinum) had been tested with negative or indifferent results. Then (26 October 1953) Breil came to zirconium*. In the very first test, half of the ethylene reacted to form a solid mass of polymer that blocked off the catalyst from the remainder. Breil immediately made a second run in which he violated the rule that in research only one variable should be changed between successive experiments. He doubled the catalyst, added a solvent and improved the stirring. The record does not show whether this short-cut shot was made on his own initiative or under Ziegler's direction.

Reasonable or not, spontaneous or inspired, these proved to be happy multiple choices, for the next experiment was an unqualified success. Rapid and complete polymerisation occurred and the product could easily be recovered as a white powder. This material had all the characteristics of a true high polymer, and even at this stage showed properties that were visibly superior to those of any previously known polyethylene.

At this point there was no difficulty in recognising that a potentially valuable process and product had been discovered. A description by Ziegler, given fifteen years later, is still graphic:

'When we cut a small piece out of this sheet' (made by pressing the powder between metal plates in a small hot press), 'we were able to stretch this, I think, about three or four times its original length, and this is very characteristic of a certain type of polymer, polymers which can be transformed into fibres; and this stretched piece was very, very strong. It was impossible to put it into two pieces. It was impossible by my own hands to put it into two pieces. The expert [in infrared spectroscopy] declared to us that the peak which had disappeared in our spectrum . . . was the peak which demonstrated that the high-pressure polyethylene is branched; and so we came to the result that . . . we must have a really new type of a really straight-chain high-molecular-weight polyethylene.'

---

*This relatively rare metal is most commonly encountered in the form of the 'imitation diamond', zircon (a silicate).

> Breil's thesis gives a somewhat more matter-of-fact and quantitative description of his first product, but one that is impressive in its scope. He reported that it had a softening temperature of 130–150°C (266–302°F), a good 30°C higher than previously known polyethylenes, and could be pressed at its softening temperature into transparent sheets or foils which were noticeably stronger than corresponding products from high-pressure polyethylene. A strip cut from the foil could be elongated approximately four times its length, with an appearance similar to that of polyamide (nylon) stretching. The stretched film showed great strength.
>
> These properties were recognised as indicative of high molecular weight and a linear structure. Viscosity molecular weight measurements indicated a value of about 200 000. X-ray diagrams of highly oriented strip resembled those of lightly oriented high-pressure polyethylene. Infrared spectra resembled those of conventional polyethylene except for absence of an absorption band characteristic of methyl groups (formed by branching).
>
> A modern polymer laboratory could do more with much less material, but for 1953 in a classical organic chemistry laboratory, it was something of a *tour de force* to glean that amount of information and conclusions from the very first meaningful sample.

Breil's thesis displays a photograph of the powdered polymer; beside it are the crude first film pressed from the powder and a piece of tubing made by moulding a cylinder and drilling out the centre. When I again visited the Max Planck Institute twenty years later, I found that these historical samples had been carefully preserved and kept on display in a showcase.

Whether from momentum, thoroughness or the insatiable quest for the best, the testing of other metal compounds continued unabated, despite the striking results obtained with zirconium. When the turn of titanium* came up, it was tested in the, by then, standard reaction conditions: temperature 100°C (212°F); ethylene pressure of over one thousand pounds per square inch; catalyst system, aluminium triethyl and titanium tetrachloride. The result was striking. The reaction was so fast that it became very hot, and the product was so charred and partly

---

*Titanium, now well known from its use in aerospace projects, was at that time a little known element, except in the form of its oxide, used as a white pigment.

decomposed by the heat that it was scarcely identifiable as a true high polymer.

It was Heinz Martin who tamed titanium. Martin, a highly competent research chemist, had already completed his doctorate under Ziegler, studying the action of organic aluminium compounds on propylene. That he worked with propylene is intriguing in the light of later events; however, his dissertation dealt, not with polymerisation, but with one-step additions to produce a five- or six-carbon olefin (depending on whether aluminium triethyl or tripropyl was used). This process was eventually licensed to a major U.S. rubber company (Goodyear) who use it in making isoprene, the raw material for 'synthetic natural rubber'. Since the latter product was in turn made possible by 'Ziegler Chemistry', we may say again, with Ziegler, '*les extrèmes se touche*', or at least that things often come full circle.

Martin, a poised and personable man, had a command of English, a breadth of interest and a balance of outlook that were exceptional for a young German research scientist. He had benefited in some respects from an inadvertently extended sojourn in the United States. Entering as a young exchange student, he was prevented by the outbreak of World War II from returning to Germany, so he continued his education in America through high school and into college (M.I.T.). In later years he became one of the top two men at the Max Planck Institute and, upon Ziegler's semi-retirement, was put in charge of worldwide licensing of Ziegler's and the Institute's proprietary processes*.

In view of his personal attributes and the fact that he had already completed his doctoral research, it was natural for Martin to be assigned to those aspects of the program that were of technological or practical importance, rather than of purely theoretical interest. At the time of the polyethylene discovery, he was looking for reaction conditions that would drive the molecular weight to much higher levels; ironically, after the co-catalysts were found, he had to reverse his field and learn what conditions would bring the super-molecular-weight polymers they made back into the practical range.

In addition, while Breil was marching through the periodic chart, Martin was juggling catalysts and conditions in a search for a polymerisa-

---

*In order to preserve the non-profit status of the Institute, a separate company was formed to conduct the actual licensing and collect royalties from which it contributes funds to Institute support.

tion system that would work under the mildest possible conditions. The system already at hand (zirconium co-catalyst) represented a thirtyfold betterment of the superpressures required by the previously known ICI process (which made polyethylene with a branched and irregular structure); nevertheless, still gentler conditions would have further practical advantages and might make a still better product. Since it was apparent from Breil's first experiment that the titanium/aluminium alkyl system was a 'hot' one, Martin tried it under the simplest possible conditions: no pressure at all and no external heating.

The dramatic *dénouement* was witnessed by Ted Borrows of Petrochemicals Ltd, who was visiting Mülheim at the time. The visit was not casual, nor the timing accidental. Ziegler had written to Borrows soon after Breil's first successful experiment, sending him a bit of the polymer and an invitation to come to Mülheim. He indicated that the 'new development' fell outside the original agreement that he and Borrows had negotiated and that it was therefore necessary to arrange an extension of the agreement that would give PCL rights to this important breakthrough. Borrows, whose German is much better than that of most Englishmen, nevertheless had the letter translated three times by experts to be certain of the exact message. Then he took the first available plane to Mülheim. His mission was to persuade Ziegler that no new licence (or fee!) was necessary because the wording they had jointly worked out for the original licence would apply to the new finding as well.

His position was compounded from equal parts of optimism and bluff, since PCL had no information at that point on how the new results had been achieved. They had a 'man in residence' in Ziegler's laboratory by an arrangement similar to Natta's, but he did not cultivate sources of information as assiduously as did Natta's 'Three Bright Boys', and so did not learn of the discovery any sooner than Borrows. So it came about that Borrows spent a full and hard day in Mülheim going round in circles with Ziegler.

'It is outside the agreement', Ziegler kept insisting.

'I can't agree to that', Borrows maintained, 'if I don't know what I'm agreeing to.'

Neither was apparently familiar with, or inclined to use, the mechanism for confidential disclosure under secrecy agreement that is now commonplace in licensing negotiations, but they arrived at an informal equivalent when Ziegler fell back on the 'gentlemen's agreement' they had previously enjoyed.

'Dr Borrows, I trust you', he said finally, 'so I shall tell you what it is in confidence.' He then disclosed the secret of the co-catalyst.

'I'm sorry, Prof. Ziegler', Borrows said immediately, 'but then it's within the agreement. Let's go through the wording once more.'

Typically, Ziegler did not refer the question to legal counsel for a professional opinion at this point. Instead, after poring meticulously over the language before them, he simply acknowledged:

'You are right, Dr Borrows!'

It was very shortly thereafter, while Borrows was still in Ziegler's office, that Heinz Martin burst in–an unusual familiarity–waving a glass flask and obviously excited.

'*Es geht in Glas*! It goes in glass!', he cried.

He proudly displayed the white powdery polyethylene he had made under his no-pressure, no-heat conditions.

Had this bit of drama occurred somewhat earlier, Borrows would have suspected that it was being staged for his benefit, but under the circumstances he could only presume that he was witnessing another major breakthrough. He was quickly convinced of that by what followed. Karl Ziegler was not the demonstrative type, nor was his Institute noted for frivolities, but the Director immediately decided that this triumph warranted a celebration. He graciously invited his visitor to join in the festivities, but excused himself on grounds of being older and a bit tired. He sent Martin and Borrows off with two other senior staff members (Wilke and Zosel) to the nearest spot with any real 'night life', Dusseldorf. As they left, Borrows told Ziegler he would see him about nine o'clock the next morning to wrap up their discussion. Ziegler smiled.

'Dr Borrows', he said, 'I'll see you when you can make it.'

Borrows 'made it' barely by noon, after a night on the town that lasted until four in the morning.

The 'golden time'

The next few months must have been a 'golden time' for Ziegler and his colleagues at the Max Planck Institute–a time when they could share both a sense of achievement and a keen anticipation of things to come–in many ways, a better time than later days that were filled with fame, fortune and frustrations. They held in their hands a resounding research success that offered every promise of scientific recognition, additional exciting research leads and rewarding practical applications. Ironically, their pursuit and enjoyment of these first fruits were to suffer from the very fact that the scope and importance of their new reaction system proved to be still greater than they imagined. What made that upsetting was that much of the credit and attendant publicity (and remuneration!) for the additional discoveries was garnered by others, sometimes justly and with proper acknowledgment, sometimes not.

The Phillips discovery

The laboratory in which Phillips polyethylene was discovered is a long way from Ziegler's laboratory, both in miles and in environment. Bartlesville, Oklahoma, halfway between Herd and Hog Shooter, was born out of the oil fields of the Oklahoma Territory only months before Karl Ziegler's birth. It grew into a prosperous modern town, thanks largely to Frank Phillips and those same oil pools. Bartlesville is inescapably an oil industry–and oil company–dominated town, however modern, and Phillips Petroleum Company is inescapably paternalistic, however benevolently.

Phillips has a record and a reputation which are impressive, at least among oil companies, for venturing boldly into new business areas and for being something of a maverick in its business approaches. However, it is highly orthodox in the degree of company loyalty expected (and exacted) from its employees. Technical successes have won for research internal respect and strong support; however, the research is frankly 'mission oriented'. The laboratory staff, by and large, have been chosen and conditioned for interest in practical fruition and ultimate payout to Phillips for the results of their theoretical studies*.

---

*Purists contend that research that has an identifiable goal is not 'pure' and hence is unworthy of being called research. They, in turn, can be accused of scientific bigotry, or at least of specious semantics, for who among us would maintain that his research was utterly aimless? The true difference lies in how well the goal, the means to it and the odds of reaching it are acknowledged and specified, and in whether the benefit sought is primarily for the researcher himself, a sponsor or some larger segment of mankind. In some happy instances, such as the one at hand, all three are beneficiaries.

Hogan, hero

John Paul Hogan is, by his own description, so oriented. Born near Paducah, Kentucky, he graduated from that state's Murray State College in 1942 and spent the war years teaching chemistry to high-school students and physics to Air Force Cadets at Oklahoma University. After the war he stayed in Oklahoma, joining Phillips as a research chemist. A few years later, he and R. L. Banks were assigned to a research project with the specific goal of making gasoline-range hydrocarbons from petroleum gases. As a general background, they had a wealth of previous knowledge to draw on. As usual, some of that knowledge originated in Germany.

As was the case in Ziegler's laboratory, the seminal 'prediscovery' was not of a polymer at all, but simply dimer. In 1932, a German chemist named Schuster discovered a fairly selective catalyst–cobalt on charcoal–for converting ethylene to dimer (1-butene and 2-butene). This was another accidental discovery; Schuster was investigating the hydrogenation of ethylene.

During World War II, when butene was in desperate demand for conversion to butadiene and thence to synthetic rubber, this reaction was developed into a continuous process by research teams at Shell Development Company in California. Reports of their work were circulated to other oil companies in the government-sponsored program, and patents were issued in 1946. The Phillips group doubtless had access to this literature and information; they also had an in-house discovery by Bailey and Reid (patented in 1945) that ethylene could be converted to liquid polymers using a nickel oxide (on silica-alumina support) catalyst.

Hogan and Banks developed a modification of the Bailey-Reid process to make dimers and trimers from both ethylene and propylene, but they were bothered by the short life of the catalyst. They tried various additives to confer longevity, but never found the fountain of catalytic youth. However, they made one key experiment in which propylene gas was passed through a one-inch diameter tube packed with catalyst to which a chromium salt had been added. This experiment failed, because the tube quickly became plugged with a solid.

This was not the first time such a result had been observed and reported. In the wartime work at Shell Development Company, attempts had been made to go directly from ethylene to butadiene, using cobalt and, in one case, chromium, as catalysts. Difficulties were repeatedly encountered with reactor plugging by 'waxy solids'. Fixed by war-driven necessity to a rigid goal, the Shell workers regarded the solid polymer as a nuisance and the experiment as a failure. They made no investigation of the nature of the troublesome polymer, and wrote an epitaph to this line of research:

'Use of inert diluent did not prevent the accumulation of high-molecular-weight polymers including some which formed waxlike solids at room temperature. Loss of catalyst activity accompanied the appearance of these heavy polymers . . . this work was discontinued.'

In the post-war era, research was no longer so tightly regimented, and at Phillips a similar result got dissimilar treatment. Even though their experiment was a failure in terms of their original objective, Hogan and Banks did not discard their plugged pipe; instead, they examined it to see what had happened.

What had happened, of course, was that they had made a (relatively) high polymer of propylene. This initial product was not very exciting, but additional experiments using ethylene gave a more promising material. Even so, Phillips management might well have felt that pushing this unproven new ethylene polymer into an unfamiliar field dominated by the mighty duPont Company would be a hard and risky course (as, indeed, it proved to be) and that, as an oil company, they should stick with their original goal of making gasoline. But Phillips had bearded lions before—notably, when they took on the Cabots of Boston (Cabot Carbon Co.) with a new type of carbon black made from oil instead of gas and converted the whole industry to that base—and they did not allow the reasonable, conservative view to stifle enquiry in this instance either. Instead, Hogan and Banks were permitted to take off on this interesting bypath that diverged from their main research road. But it was no path of roses; in Hogan's words, the project 'had to fight its way from a beginning that could be described as a few grammes of a brittle plastic made in a hopelessly inefficient process to its present status as the leading process for production of linear polyethylene.'

Phil Arnold, Vice President for Research and Development, explained management's options metaphorically:

'It does not usually pay to be distracted by rabbit tracks if you are hunting deer. But if it is getting late in the day, you are hungry and you need something for dinner, a rabbit in the pot may be better than a deer in the woods.'

Hogan preferred to believe that they had been hunting rabbits and had bagged a deer. But, deer or rabbit, it was still a long way from the woods to the pot. Phillips management had really derived their boldness from the strength of their raw material position and their engineering capability, abetted by a happy ignorance of the pitfalls of the plastics business and by the brash confidence of their marketing organisation ('We could sell snowballs to Eskimos!'). Those assets and that confidence were all to be severely tested before the new baby plastic born from Hogan's 'happy accident' would surmount its childhood diseases, pimply adolescence and youthful indiscretions and make it into the adult world of big-time plastics.

Paul Hogan, who followed his polymeric offspring through the traumatic development stages, takes justifiable pride in the ultimate achievement but is careful to extend due credit to his associates and Phillips management, starting with his supervisor, Dr Alfred Clark. He is also at pains to reaffirm his faith in 'mission oriented' research that promises a payout, but his personal concerns go beyond the stockholders' profits. Deeply involved himself with religious and community activities, he now urges his profession and his company to 'rise above a materialistic approach' and become involved with ecological problems, including, appropriately enough, the recycling of plastics.

Standard of Indiana: number two doesn't try as hard

By another striking parallelism of timing and result, the same original catalyst system studied at Phillips–cobalt on charcoal–was being investigated simultaneously at still another oil company research laboratory, although for a different reaction. This work led to still another accidental discovery of linear polyethylene.

At the Standard Oil Company of Indiana laboratories in Whiting, Indiana, Dr Alex Zletz was attempting to use cobalt on charcoal to catalyse alkylation reactions with ethylene. To his surprise, he found that considerable solid polymer was formed. Like Hogan of Phillips, he was not experienced in polymer chemistry, but had enough curiosity, initiative and freedom to pursue the intriguing bypath. Other metals and metal oxides on other supports were investigated by Zletz and his co-workers, and they eventually found improved catalysts such as molybdenum on alumina that permitted the development of a practical low-pressure polyethylene process.

Unlike Phillips, Standard of Indiana could not make up its corporate mind to take the plunge into the unfamiliar seas of the polymer business, but instead offered the process for licence. However, Phillips was also pursuing an aggressive licensing policy, supported in their case by process information developed in their own plants. Standard could not match this in-house expertise and never matched Phillips' success in licensing, but at least one commercial plant was built by a licensee (in Japan).

Thus, the same initial observation made in three separate industrial laboratories led in one case to nothing, in a second case to a limited commercial success viá licensing, and in the third case to an outstanding worldwide success in both production and licensing. The difference in outcome may clearly be attributed primarily to the different 'climates' (both external and internal) within which the unexpected observations were made.

6 Harvesting the Fruits of Innovation — Polyethylene

> There has been no explosion of laboratory activity and publications in organic chemistry on an international scale approaching that which followed . . .—Calvin Schildknecht

Karl Ziegler demonstrated how far he was from being an ivory tower academic by how promptly and aggressively he pursued useful and profitable employment for his brainchildren. We have already seen (Ch. 3) from his action after the discovery of the 'growth reaction' that he was well aware that patents are important and that tight agreements, like tight fences, make good neighbours.

We also noted that his self-confidence led him to draft his own licence agreements, with good success, and his own patent applications. He continued this do-it-yourself approach in covering the discovery of linear polyethylene. But in law, as in golf, the amateur may win some early rounds but will usually lose out to the professional in the end. So it was with Ziegler. He had told Petrochemicals Ltd that they would need to negotiate a new licence for polyethylene, but after prolonged argument with Ted Borrows he conceded that the wording of the original agreement—wording which Ziegler had concurred in—was of such breadth as to cover polyethylene also.

Much more serious was the reverse sort of error Ziegler made when he wrote the patent application so narrowly that it covered only polyethylene; but the grave consequences of that mistake did not become evident until later. Even when the attorney he belatedly engaged urged that the patent be broadened to cover other olefins, Ziegler dismissed the suggestion with the statement: 'Nonsense! Other olefins do not work!'

The news of his polyethylene discovery had set off a tide of comings and goings that swelled to an unprecedented flood, as the word spread around the world. It spread through three channels: (1) lectures and papers by Ziegler and co-workers at scientific meetings, (2) lines of communication within the chemical industry and its trade press, and (3) individual consultants and others who made it their business to keep abreast of technical developments.

Far and away the most effective of these latter was Herman Mark. He visited his friend Karl Ziegler again just after 'nickel effect' had been discovered, and he was among the first outside Germany to learn about linear polyethylene. He was also one of the very first to appreciate its potential significance. With his usual infectious enthusiasm, Mark spread the word with telling impact among the industrial and academic laboratories he visited. The ensuing scientific explosion, noted by Schildknecht in the quotation at the head of this chapter, was accompanied by, and largely fuelled by, an equally *brisant* explosion of business interest.

Many other scientists have been, and had, good publicists, but few, if any, have ever experienced anything like the overwhelming responses that greeted Ziegler. Why this unprecedented, untempered enthusiasm for an unproven laboratory curiosity?

The answer is that Karl Ziegler had the great good fortune to make one of the rare discoveries that industry can immediately appreciate, and to make it at one of the rare times when industry is prepared to make heavy commitments to launch new products and create new markets. Granted, most Ziegler licensees underestimated the problems they would face, and solve, before the new product would be successful; nevertheless, they plunged in with an enthusiasm that was infectious and thus autocatalytic, and with a resolution that carried most of them through the difficulties.

Pundits were as fond then as now of saying that 'the day of the individual inventor is past', and that the scientist who discovers something in his laboratory and promptly sells it for a million dollars exists only in movies and science fiction. Yet Ziegler did just that—not once, but many times over.

We who were self-taught 'expert' technologists and self-styled 'market analysts' were also prone then, as now, to assure one another sadly that there were no more big worlds to conquer in polymers, because there was no 'room' for another large-volume, general-purpose plastic. The flood of products released after World War II had given us polystyrene (hard, clear), high-pressure polyethylene (soft, flexible) and polyvinyl chloride (from rigid to highly flexible, as desired; also fire retardant)–all inexpenstive, all easily fabricated. What else was needed or could be found that would be cheaper or better?

Nevertheless, once it existed, the potential virtues of linear polyethylene were easy for even experts to see and exciting for businessmen to contemplate. Harder, stronger and with a higher melting point than the familiar and successful high-pressure polyethylene, it seemed obvious that it would also be easier and cheaper to manufacture. The high-pressure process was notorious for requiring highly specialised equipment and for operating at the unfamiliar and frightening outer limits of practical conditions. Runaway reactions, leaks and explosions were not uncommon. Moreover, rights to the process were held tightly by duPont and ICI. By contrast, the spectacle of solid, white polyethylene powder forming before one's eyes in a glass flask that was barely warm and under no pressure had an irresistible attraction. Ziegler's process looked appealingly (and deceptively) simple.

It was an advantage, not a drawback, that the new product was, in fact, only an improved form of a familiar one. It was still polyethylene, and everyone, even vice presidents, knew how useful polyethylene was and could see that a harder, tougher variety would be a highly saleable commodity. Marketing executives were particularly entranced. Press the average salesman hard as to what he really wants from research, and it will emerge that his first choice is to have the same product as his competitor, that his second choice is an improved grade of a known product, and that either is far preferable to a really new product that requires creating new markets and educating himself and his customers in its uses and idiosyncracies. Best of all, of course, would be the improved product at a better price, and this was just the boon that linear, low-pressure polyethylene seemed to promise.

The other fortunate factor was timing. The 1950s were a time of almost explosive expansion in the chemical industry–an unusual time when the only heinous errors were those of omission. In most companies, the annual review of the year's activities brought regrets and recriminations only for any failure to pursue an opportunity. The competitive struggle, in contrast to today's defensive strategies, was played like the Japanese game of 'Go', with all the emphasis on striking out aggressively to capture new territory and getting there before one's competitor.

In this climate, Ziegler's discovery made exciting news that called for prompt action. It offered companies already in plastics a chance to expand into a new field and those outside the fence a chance to break into the green and glamorous pastures of the plastics business. And it was obvious that the race would go to the swift.

Hercules, the hesitant pioneer

This combination of circumstances quickly made Mülheim the new Mecca for academic and industrial pilgrims of all stripes—the curious, the desirous and the determined—flocking to see, to admire and to negotiate. But they found that companies with good 'listening posts' in Europe had been there ahead of them.

As we have seen, PCL already had an exclusive licence for England, Montecatini for Italy, and Hercules (under option) for the U.S. Ziegler had also licensed Hoechst in Germany, to whom he had sent moulded samples of polyethylene in November 1953. Hoechst responded positively and became the first commercial producer of Ziegler polyethylene; however, theirs was not an exclusive, but a 'semi-exclusive', licence, since the German coal companies, as sponsors of the Max Planck Institute, were also entitled to licences.

In the U.S., however, the door did not stay permanently closed. Although Hercules had taken advantage of their early information to obtain an option on an exclusive licence for the U.S. (Ch. 4), key management people blocked a commitment to pursue that opportunity aggressively, despite urging by their own people and even by outsiders.

During the period that Hercules' option was open (September 1953), Sir Robert Robinson happened to visit the U.S. and was entertained for lunch by Hercules executives in the Brandywine Room of the duPont Hotel in Wilmington. The talk turned to Ziegler chemistry, and Sir Robert spoke with great conviction of its potential importance.

But neither Sir Robert's gratuitous advice nor the urgings of Wiggam and others within the company moved Hercules management to act; at the expiration of their option (end of September 1953) they still had not made a decision to exercise the option and take a licence. Even Paul Johnstone, usually a positive thinker and doer, acknowledged that he had a 'mostly negative' attitude toward Ziegler's process.

One month after the official expiration date of Hercules' option, linear polyethylene was discovered in Ziegler's laboratory.

Even then, all was not lost, for Hercules was given another chance. Ziegler voluntarily kept the option open as a courtesy to the Hercules men who had become his friends, this despite the fact that others were now knocking on his door. To at least one major U.S. chemical company (Union Carbide) who had sent a vice president as an emissary to Mülheim, Ziegler said that he still felt obligated to 'another American company' and therefore could not discuss a licence.

A contributory factor may have been the fact that Ziegler had agreed to let Arthur Glasebrook come to Mülheim for an extended visit to gather as much information as possible that would show Hercules the merits of the new chemistry. Glasebrook stayed at Mülheim from May 1953, until July, when the entire Institute closed down for the summer vacation. He won Ziegler's confidence*, and was given direct access to the research staff. He spent considerable time with the investigators in each of the main branches of the research program, including Gellert's work on ethylene polymers.

During this period, Glasebrook encountered several others who were there on similar missions from other countries. Staying at the same hotel were the three young Italian assignees from Natta's laboratory (Crespi, Chini and Magri; Ch. 4), and he became good friends with them. ('A very friendly fellow', one of them said of him later.) He also met the resident representative of the German firm, Hoechst, and of Petrochemicals Ltd, as well as Ted Borrows, a frequent visitor during those days when all paths meandered toward Mülheim.

The little group of outside observers, despite their diversity of background and interest, became so well acquainted with one another that they soon formed a sort of informal luncheon club that met at a local restaurant and freely exchanged information and comments.

It thus happened that Arthur Glasebrook was at the laboratory when Holzkamp made his first crude product containing some high-

---

*On Glasebrook's first day at Mülheim, Ziegler made a point of reminding him of the European tradition that when a professor opens a new area, others recognise his right to develop and extend that field for a time. He felt sure that the other observers, being Europeans, would respect that tradition, but he expressed concern that Americans might not, and that they would 'barge in' immediately. In later years, Ziegler remarked that all his difficulties in that regard had actually involved his European, rather than American, licensees.

molecular-weight polyethylene. However, his good fortune and prompt follow-up were of no immediate avail, for although he sent and carried word of this discovery back to Wilmington, it made no immediate impact there.

Back in the U.S., in December 1953, Glasebrook received a Christmas card from Magri, who sent friendly greetings and added a note about the successful synthesis of high polymer from ethylene (presumably this was from Breil's experiment with zirconium or a later one with titanium co-catalyst).

Ziegler himself wrote to Wiggam of Hercules (2 December 1953) to give him official word of the new product. On 22 December, even though Hercules' option had long since expired (September 1953), he sent Wiggam a moulded sample of polyethylene bearing the imprint of the German eagle (it had been hot-pressed against a German mark). He did not disclose the catalyst system used, and Wiggam issued instructions that the Hercules chemists were to make physical and mechanical tests only, no chemical analyses that would reveal catalyst residues.

This concrete evidence at last stirred Hercules management sufficiently to resume negotiations, but not sufficiently to accept the terms asked by Ziegler. Meanwhile (May 1954), another American firm, Goodrich-Gulf, learned of Ziegler's discovery through their own contacts in Germany, and within two months they had reached an agreement on terms with Ziegler. At this point, Ziegler finally gave up on Hercules and signed a non-exclusive licence option with Goodrich-Gulf. What they did with that option, in marked contrast to Hercules, enters our story later.

Thus, by the time Hercules finally was prepared to come to terms, only a non-exclusive licence was available, and they ended up paying more for that than exclusive rights would have cost them earlier.

Ziegler, who still felt some indebtedness to his friends in Hercules, granted them one small exclusive—the right to both make and sell his catalysts in the U.S. (others could manufacture for their own use, but only Hercules could also manufacture for sale to others*). Moreover,

---

*Hercules later formed a joint company (Texas Alkyls) with Stauffer Chemical Company to make and sell aluminium alkyls for catalytic use. The exclusivity was subsequently lost as a result of anti-trust actions.

Wiggam had foresightedly seen to it that the licence covered all aspects of Ziegler chemistry, including reactions of other olefins besides ethylene.

Latter-day polyethylene pilgrimages

As the word spread and subsequent discoveries broadened the scope of Ziegler chemistry and enlarged its industrial potential, other companies continued to send missions to Mülheim for several years. At times, the traffic became so heavy that it required tight scheduling. Some of the missions, including one of which I was a member, put themselves up at the Petershof, a castle-like hostelry on top of the Petersberg, overlooking and across the Rhine from Bonn. The splendour of its view and services, combined with its historical fame*, created an amply impressive setting. While our delegation was conferring in the lounge on the eve of our appointment, we recognised in another group across the room individuals, who, we knew, could be there only on the same mission for another company. The dimensions of the spacious lounge made it unnecessary for the two groups to acknowledge each other's presence, but the unvoiced competitive message was clear.

Ziegler's efficient secretary had scheduled the next day's meetings so that the rival groups did not encounter one another. Mülheim is a surprisingly attractive city, considering its location in the industrial Ruhr valley, and the Max Planck Institute is situated along quiet, tree-lined streets in a residential section. In that setting, the long, ivy-covered brick building was reminiscent more of a university campus than an industrial institute. Inside, the illusion persisted as the visitors were ushered into Prof. Ziegler's book-lined library-cum-office and seated at a low, round table.

Ziegler now had a professional attorney as his patent and licensing counsel: Dr Andreas von Kreisler. After introductions, Ziegler explained that he, a research scientist, was interested primarily in the scientific aspects of his work and had delegated to Dr von Kreisler responsibility for settling business matters. But instead of excusing himself at that

---

*Neville Chamberlain stayed at the Petershof when he came to Germany for a fateful pre-Munich meeting with Hitler, and General Eisenhower used it as his headquarters at the close of World War II.

point, he went on to describe the nature of the licences that were available. When von Kreisler finally got the floor, he carefully prefaced all his statements by saying:

'Assuming Prof. Ziegler agrees, . . . ' or,

'I have suggested to Prof. Ziegler that he might consider . . . '

His deference was not a pose; von Kreisler knew his man. Every few minutes, Ziegler would break in to emphasise a point or express his own opinion, and he did not hesitate to interrupt or correct his counsel with a 'Nein, nein!'. It was never in doubt that the chemist, not the lawyer, was running the show. Nevertheless, it was clear that both men had great respect for each other.

Ziegler's insistence on such direct participation, and his practice of hard but fair bargaining stemmed largely from an earlier disillusionment. In pre-Mülheim days he had discovered a highly useful chemical reaction (halogenation of unsaturated compounds with N-bromosuccinimide) and had been rewarded professionally with peer recognition but not financially. His name was given to the reaction (the 'Wohl–Ziegler Reaction') but it did not appear on a patent, so none of the chemical and pharmaceutical companies who used the process commercially ever offered him any compensation or even recognition. Thereafter, Ziegler determined that this oversight would never be repeated.

Between negotiating sessions, VIP visitors, lecture appointments* and publications, Ziegler could no longer find the time to discharge his regular duties and maintain his accustomed intimate contact with the day-to-day research activities. He reluctantly reduced the time he spent in the laboratory and willingly delegated more administrative detail to Wilke and others. Some of the research staff had to stage so many demonstrations of ethylene polymerisation for visitors that the progress of their research suffered.

But it was all worthwhile in a game where the rewards were so high. The price of a licence (non-exclusive) varied with the time and circumstances, but there came a time when a purchase price of a million dollars was not unusual, with running royalties to follow. Initially, the licence income went entirely to Ziegler, with smaller shares to the co-workers

---

*He had already been guest professor at Chicago (1952) and was also visiting professor at the University of Illinois and the University of Wisconsin.

he designated. But he did not let sudden affluence greatly alter his life style*. Nor did he entirely neglect his Institute's needs; at least one American licensee (Goodrich) was asked to make payment for their option, not in money but in its equivalent value in laboratory instruments, sorely needed and hard to come by in post-war Germany.

He did use some of the newly found wealth to build a new house for himself and his family, thus freeing the quarters previously occupied in the laboratory building. But the millionaire did not move to the suburbs; he located his new house directly across the street from the Institute!

Petrochemicals Ltd

PCL, one of Ziegler's first licensees, continued to enjoy a special position. Sir Robert Robinson accompanied Ted Borrows to Mülheim in the spring of 1955, and the eminent English and German scientists developed an immediate rapport based on professional respect and common interests in music and mountain climbing, as well as organic chemistry. Ziegler found that Sir Robert could speak as fondly and familiarly as he of the Swiss peaks and passes, and thereafter Borrows, who had just introduced them, felt like the odd man of the trio.

In order to inaugurate a regular flow of information from Ziegler's laboratory to PCL (for Ziegler had no formal reporting system other than his chemists' laboratory notebooks), it had been arranged for PCL to 'second' one of their younger chemists to work in Ziegler's laboratory and keep PCL up to date by correspondence. Their first 'man in residence', Louis Bohm, was at Mülheim contemporaneously with Montecatini's 'Three Bright Boys'. However, he made no effort to ferret out and pass back information that did not come his way naturally; in fact, Borrows first learned about the polyethylene discovery directly from Ziegler before Bohm had even heard of it.

Later on, Bohm was replaced by E. W. Duck, who did his thesis work at the Max Planck Institute, took his doctorate under Ziegler,

---

*Not so some other recipients. Holzkamp, for one, dropped out of chemistry as soon as his royalties began mounting up.

married a German girl, and thus became very much a member of the Institute 'family'. Many years later, Duck became technical director of the International Synthetic Rubber Company Ltd, in England.

The laboratory and pilot plant crews of PCL were meanwhile struggling valiantly to meet their commitments of producing a saleable Ziegler polyethylene on a substantial development scale. As usual, costs and difficulties greatly exceeded expectations, and PCL's board began to suspect that they had swallowed more than they could digest. Karl Ziegler, being well disposed toward his early licensee, at one point offered to extend their exclusive licence to include the entire British Commonwealth for a mere additional £25 000, but the PCL board could not be persuaded to raise their ante by even that much. When this was sadly relayed to Ziegler by the PCL representative, he said, with a consoling twinkle: 'Never mind, then I shall give you Trinidad for nothing!'

The fruitful, frantic fifties

By the mid-1950s, the surge of research, process development, and product application work began in earnest in the laboratories of Ziegler licensees, in many university laboratories, and in still other industrial laboratories where it was hoped something could be found that would make a Ziegler licence unnecessary or less expensive. Those in the universities found that they had a happy hunting ground among the problems of reaction mechanisms, catalyst structure and polymer physics associated with Ziegler polymerisation. The industrial licensees found that they, too, had bought a 'hunting licence', as one of them put it. In their million-dollar package they found nothing an engineer would call process information, nothing a rheologist or product application expert would call adequate polymer characterisation or application data, nothing a lawyer or librarian would call systematic documentation.

'What we got', said one of the recipients, 'was laboratory notebooks.'

It was only a slight exaggeration; laboratory information was all that had been offered. Everything else needed to build a commercial plant had to be created by the licensee's own chemists, engineers and physicists. Since each company was on its own, except for a certain number of feedback and exchange agreements, this made for a great deal of duplication, but also for a great deal of additional innovation. The yield was innumerable catalyst and process variations, thousands of patent applications, tens of thousands of pages of publications and ten or more commercial plants in several different countries.

The labours of Hercules (Inc.)

Hercules, despite their lengthy dalliance, were the front-runners for Ziegler polyethylene in the U.S., and they experienced all of the difficulties and some of the rewards that are the lot of the pioneer. When they finally signed their licence agreement with Ziegler, a team had been promptly despatched to Mülheim to collect all available information. Recognising the limitations of the technical data, they had also (in 1955) made a licence and information exchange agreement with Hoechst, the first Ziegler polyethylene producer on the Continent. This latter move saved Hercules an estimated two years in getting a plant on-stream*.

When the momentous day arrived (18 June 1957) to dedicate the first Ziegler plant in the U.S., Hercules arranged to bring it off with flair. They invited Karl Ziegler to attend the ceremonies, make a speech and turn the valve that would admit the ethylene feed to the reactors for the first time.

Things went according to plan—almost. Ziegler came, and made a few complimentary remarks appropriate to the occasion. He characterised the plant opening as 'one of the most significant days of my life' and his invention as 'the fruit of a long, systematic research in the field of organometallic compounds, but nevertheless it was finally discovered by chance.'

Then he turned the valve, but this action could only be symbolic. For the supply of ethylene had been contracted from the huge Standard Oil Company of New Jersey refinery some miles away, and unfortunately

---

*The information exchange also had further advantages to both parties, because its scope had been defined sufficiently broadly ('The Polyolefin Field') that when polypropylene was discovered it came under the same umbrella. Hercules and Hoechst made separate licensing deals with Montecatini, but exchanged their own process and product development information with each other rather than with Montecatini. Hoechst made another agreement with Imperial Chemical Industries in England, who eventually became the most aggressive developer of polypropylene for fibre applications.

Standard had not met their commitment regarding date of availability. So the actual plant start-up had to be postponed; nevertheless, the dedication was widely and properly reported as an historic event—one of the rare occasions when an inventor not only sees but presides at the commercial fruition of his discovery.

Standard Oil Co. (N.J.)-Esso-Enjay-Humble-Exxon

When the U.S. Supreme Court, in 1911, forced the break-up of the original Standard Oil Company (N.J.) (formerly Standard Oil Company of New Jersey, formerly The Standard Oil Trust, formerly Standard Oil Company), the name, 'Standard Oil Co. (N.J.)', was retained by the largest of the thirty-four corporate fragments. An observer who noted all of its subsequent name changes might conclude that the company must suffer a chronic crisis of identity. Whether helped or hindered by its multinominality, the company now known as 'Exxon'* has again become the country's largest oil company and has regained top rank among the world's giant corporations. Along the way, it played a much more significant role in the story of polyolefins than the negative one related above. Since the name, 'Standard Oil Co. (N.J.)', was in effect at the start of our story, and in order to avoid tripping over successor and subsidiary names, we shall refer to them generally as 'Standard'.

As the original Standard Oil Company had pioneered in creating the oil business, so Standard Oil Co. (N.J.) were pioneers (although this time not alone) in creating the petrochemical business (chemicals from petroleum). Part of its success in that field stemmed from an agreement it had made in 1929 with I. G. Farbenindustrie, the giant German chemical company, by which it obtained information and rights to certain technical processes. This was followed a year later by the formation of a joint company to develop new chemical processes. (I.G.F. sought

---

*In a final (?) move to free itself from the confusion and connotations of the various 'Standard' names ('Esso' = S.O.; 'Enjay' = N.J.; there were also six other independent Standard Oil Companies in various other states) and the obviously inappropriate connotations of 'Humble', the fabricated name, 'Exxon', synthesised with the aid of a computer, was chosen as one having no roots in either corporate or human history. As we noted earlier in the case of the corporate initial fad (Ch. 2), many corporate names now seem designed to foster, rather than combat, the image of the 'faceless' and 'soul-less' corporation.

thereby to strengthen its raw material position; Standard to gain new technology. On the record, Standard (and the American people) gained more from the agreement than did the Germans.)

But these arrangements were broken up (as was I.G.F.) by World War II, and Standard was enjoined from renewing similar ties with any of I.G.F.'s post-war fragments.

Thus, when other American companies were establishing contacts and 'listening posts' in post-war Germany, which they rightly perceived as still a potentially fertile ground for new chemical discoveries, Standard was self-consciously avoiding any too-entangling foreign commitments. Moreover, there was a feeling within Standard that Germany was finished and that it would be England that would probably provide the technological leadership in post-war Europe.

And so, ironically, when Ziegler's key discoveries were first becoming known through various unofficial channels, Standard not only enjoyed no advantage from its former close ties with the German chemical industry, but was actually at a disadvantage with respect to its U.S. competition. As we have noted (Ch. 3), Ziegler lectured extensively in the U.S. in 1952 and 1953, and Standard had an opportunity to invite him for a lecture to the company's research staff, but passed. (Hercules, with the same opportunity, took advantage of it, with positive and far-reaching results.)

The end result was that Standard, far from being the first, was actually quite late in taking an active interest in Ziegler chemistry. But the early contacts with I.G.F. did have one lasting result that gave Standard a real advantage once they became involved. That advantage was a familiarity with high polymers, particularly polyolefins.

One of the discoveries made in the I.G.F. laboratories and passed on to Standard was polyisobutylene, a rubbery gum of modest practical utility*. Standard began investigating this polymer in its own labora-

---

*It is of at least historical interest that polyisobutylene, which today is not even mentioned in most discussions of polyolefins, was actually the first one discovered (as in so many other instances, the discovery was accidental) and the first one for which a helical structure was proposed (*see* Ch. 9). It was also the first linear polyolefin to be produced commercially. Moreover, studies of this polymer contributed much to the early development of the science of polymer physics, as well as leading directly to the discovery of butyl rubber.

tories, and this led to the discovery of a more valuable polymer. In the late 1930s, R. N. Thomas and W. J. Sparks found that by mixing isobutylene monomer with a little isoprene (to provide some unsaturation; Ch. 10), they could create a co-polymer that was a vulcanisable, and therefore more useful, synthetic rubber. Standard went into the commercial production of this polymer in 1943, with some considerable benefit to the war effort*. They christened it 'Butyl Rubber'.

Also investigated were the German 'Buna' synthetic rubbers (from *bu*tadiene and *na*trium (sodium), the catalyst originally used in the polymerisation). Although Americanised versions of 'Buna S', a butadiene-styrene co-polymer, became the world's foremost tyre rubber, Standard never undertook its manufacture, reputedly from lack of enthusiasm for competing with the major tyre companies. They did manufacture 'Buna N' (butadiene-acrylonitrile co-polymer), a speciality oil-resistant synthetic rubber, but later sold that business.

It is significant also that Standard already had a program directed to making polyethylene—by the original high-pressure process. The first work had been done by L. Eby and D. L. Cottle, under R. N. Thomas, and the program had advanced to the pilot plant stage. There was also even a research effort (under H. Wilson) aimed at modifying the process to produce a more linear polymer.

Finally, and very importantly, Standard's management happened to be in a positive-thinking phase in which chemicals and, particularly, polymers were looked upon as attractive ventures.

Thus, Standard was far better prepared in terms of research capability, commercial experience and management receptivity than many other oil and chemical companies that became intrigued with Ziegler's and Natta's processes. The inventors of butyl rubber were still in the laboratory (perhaps not fully satisfied with the recognition they had received, but not dissatisfied enough to leave), and they figured in the program that was soon to be launched on stereoregular polyolefins.

Even though Standard had sat out the preliminary rounds of Ziegler chemistry, when the main event came on (the discovery of linear polyethylene), they were actually only months behind other American com-

---

*Butyl's big assets are low gas permeability and excellent weather resistance. Its biggest uses are in inner tubes, sealants for tubeless tyres, gaskets, caulkants, etc.

panies in undertaking serious development work. Despite their 'stand-offish' position in Europe, they did have some liaison contacts there. Via these, via publications and via what one of their people called 'seepage', enough was learned by the autumn of 1954 to convince management that Ziegler chemistry might present important business opportunities. A team* was therefore despatched to meet Ziegler and Martin in Mülheim and discuss licensing terms. A sample of Ziegler's polyethylene was obtained, with (surprisingly) no restrictions on what might be done with it.

This sample was turned over to a research group headed by R. N. Thomas, of butyl rubber fame. They worked swiftly. Analysis for catalyst residues showed aluminium (and titanium?), suggesting that the process was related to Ziegler's already published 'growth reaction' (Ch. 3). By Christmas, J. Lippincott, working under Thomas, had done enough experimental work to make it fairly clear what the new catalyst system must be.

The chemists at Standard probably felt at first that they had a lot of ground to make up. Not only did several competitors have a head start, but one of them, Hercules, had what looked like a formidable advantage in having access to process information from Hoechst, the German chemical company that was first in the field with actual development work. Moreover, Phillips had announced their own polyethylene process (Ch. 5) and indicated that they would license others.

In actual fact, however, most of Hercules' potential advantage in time and position had been dissipated by their management's indecision. Furthermore, the people most directly involved discount the value of the Hoechst information. For one thing, the grades of polyethylene Hoechst developed for the European market were poorly suited to U.S. demand (which rates easy, fast mouldability above product quality). For another, Hoechst were so satisfied with linear polyethylene as a product that they were slow to take a real interest in other polyolefins, particularly polypropylene.

Standard, by good judgment and/or good fortune, made up for lost time by launching a strong research and development program that soon made some significant contributions. Early in 1955, William Asbury, a member of the team that made the initial contacts and who became Vice

---

*Charles Morrell, Paul Smith, William Asbury.

President for Patents and Licensing, completed negotiations for a Ziegler licence. Full technical information was then obtained and experimental work went forward on several fronts. (A total of seventy-three professionals were eventually involved in this program.)

Among those assigned was Dr Art Langer, who had taken his Ph.D. at Ohio State, had joined Standard in 1951, and was now a section leader in the research laboratory.

Langer had actually made a linear polyethylene using butyl lithium and tertiary amines as the catalyst system (compare this with Marvel's 1930 experiments; Ch.2), but that route offered no promise of economic feasibility. He now undertook (with A. R. Killeson, a Norwegian chemist) exploratory research on Ziegler-type catalysts and the effects of various additives as 'promoters', or co-catalysts.

Later in 1955, the Standard research staff was joined by Dr Erik Tornqvist, with a fresh Ph.D. from the University of Wisconsin. Trained in Sweden as a chemical engineer (cf. Natta), he took his graduate degree in biochemistry, with little attention to high polymers except those that occur in nature. But he had already heard Ziegler discuss his 'growth reaction' when the German professor lectured at Wisconsin in 1953. Tornqvist's initial assignment at Standard was on Ziegler chemistry, working on improvements in both catalyst and process.

Along with many others, Langer and Tornqvist soon became involved with other products of Ziegler/Natta chemistry, and it was in these fields that they were to make their most significant contributions. For Standard never did build a Ziegler plant (although they did become a major producer of high-pressure polyethylene), but made their commercial entry in polypropylene, as we shall see later.

Polyethylene 'goes straight'

The companies who had licensed the Phillips process were better off in regard to process problems, because Phillips had had to solve them in order to build their own plant, and they made the engineering data available to their licensees. However, everyone went through much the same traumatic adolesence in developing product application information. For the twin offspring of the 'Chain Straighteners' on both sides of the Atlantic–Ziegler's and Hogan's (Phillips') polyethylenes–turned out to be '*enfants terribles*'.

As soon as enough linear polyethylene became available to try to put it to work, it was found that the 'Chain Straighteners' had done their work too well, for in polyethylene as in society, absolute straightness turns out to be more virtue than is wanted for practical purposes. The advantages of hardness, stiffness, strength and heat resistance in finished products were gained at the cost of serious difficulties in the manufacture and use of those products. Since Phillips polyethylene had the 'purest', 100% linear chain structure, it suffered these difficulties in the highest degree. It was also the most highly crystalline.

The major developed markets for which linear polyethylene seemed suited were moulded articles, blow-moulded bottles, extruded film and extruded shapes (e.g. sheet, pipe). In every one of these fields, the initial bright promise was tarnished by one or another unanticipated 'different' characteristic that proved to be a significant disadvantage. In moulded articles and extruded sheet, the problem was excessive warpage, brought on by high and uneven shrinkage. Even more unexpected and disconcerting was the discovery that a moulding exposed to hot air for some hours would lose all its strength and practically fall apart. The pieces could be reground and remoulded to make a moulding as good as the original, but this was neither explicable nor comforting*.

The established markets for packaging film were not open to linear polyethylene because its high crystallinity rendered it opaque and prone to splitting. Blown bottles were easy to make, but out of them emerged an evil genie—one that caused the bottles to develop mysterious cracks and embarrassing leaks. Confusion about the nature of this phenomenon is revealed in the plethora of descriptive names invented and applied as a substitute for understanding: 'stress cracking', 'stress corrosion', 'detergent cracking', etc.

And, finally, even the field of extruded pipe fathered a failure that was the more galling because it sneaked up on the experts and flatly

---

*Later evidence suggested that a moulding from highly crystalline polyethylene consisted essentially of close-packed microcrystals held together by a few stretched 'tie molecules' which, thus strained and exposed, are readily broken by oxidation. The destruction of this tiny fraction of the total mass resulted in mechanical failure, but remelting and remoulding would establish new crystal domains and new 'tie molecules'. This situation was revealed in some elegant electron microscope pictures published by Keith and co-workers at the Bell Telephone laboratories. So-called 'stress corrosion' of blown containers is probably related to this same type of structure.

contradicted their predictions. Pipe made by extrusion from linear polyethylene passed all short-term tests with flying colours and seemed the ideal product for low-cost industrial and residential piping. Ziegler was thus made to look like a prophet, for the first 'shape' ever made from Ziegler polyethylene was a piece of 'pipe' a few inches long, laboriously hand-fabricated from the product of Breil's first successful experiment.

Careful measurements had shown that, under continued load, such pipe would begin to swell almost imperceptibly ('creep'), but at such an extremely slow rate that it was predicted to have a service life of at least 50–100 years. Marketing campaigns were prepared on the strength of this assurance, only to come to an embarrassed and expensive pause when it was discovered that after just a few years the 'creep' rate would suddenly accelerate and the pipe would grow blisters and eventually burst.

All these nightmarish problems came to light just as the first polyethylene plants were coming on-stream, with all the trauma associated with plant start-ups. One after another, each of the 'miracle progeny of plastics' began to look more like a retarded child. The company managements that had fathered them were forced to contemplate infanticide and were, in turn, threatened by their boards of directors with financial sterilisation. Heads were greying, if not rolling, in more than one organisation, and Phillips, for one, were very close to closing their plant and aborting the whole enterprise when salvation showed up in the form of a toy.

It would be gratifying to a chemist to be able to report that the research teams had leaped to the rescue just in the nick of time with technical solutions to all the technical problems. As a matter of fact, they did eventually solve nearly all of them, but not quickly enough to save the situation had it not been for the fortuitous burgeoning of an unprecedented, unplanned, unforeseen market. The Wham-O Toy Company introduced the 'Hula Hoop', made from extruded polyethylene tubing, and the ensuing craze created a non-critical demand that swept the glutted warehouses clean and put the plants back on an around-the-clock production schedule.

This fad, even though it faded rapidly, bought time for the polymer chemists to determine that most of their woes arose from the super-crystallinity of their super-linear polymer chains, and that by judicious adulteration with touches of a second monomer (introduced in the polymerisation step), they could put a very occasional kink in the molecular chains and thus alloy the crystallinity of the polymer just enough to suppress most of its deficiencies while preserving most of its sterling qualities. Armed with such modified products (from either the Phillips or the Ziegler process), the marketers returned to the fray, and with such success that the age of polyolefin plastics was finally firmly launched*.

The daily crises now reverted to the familiar ones of quality control and new grade development for new uses. However, before long a new question arose: how to deal with the lusty infant that had just joined the family—polypropylene.

*DuPont and ICI would probably maintain that their high-pressure polyethylene had launched the age a decade or two before; however, the class name, 'polyolefin plastics', was not applied until there was more than one member of the class. Moreover, with the added impact of the almost simultaneously spawned polypropylene, a new era can fairly be said to have begun. Unquestionably, the era of stereoregular polymers, including the stereorubbers but not high-pressure polyethylene, had its real birth in the mid-1950s.

7 Fruitful Innovation — 2. Polypropylene

The creation of a new method of obtaining polyethylene at atmospheric or slightly increased pressure is after all, in spite of its great historical and practical importance, only the perfecting and modification of a material already known in science and technology, but the preparation of a solid polypropylene involves the creation of a new, previously unknown product.—Giulio Natta

Ziegler and polypropylene

When scientists discuss the contributions made by Ziegler and his 'school', they usually start by marvelling at his discovery of linear polyethylene and end by marvelling: 'Why didn't he discover polypropylene?' The question is out of order; the fact is that polypropylene was made in Mülheim independently and contemporaneously with its discovery in Milan. Natta's recognition as the 'discoverer' of polypropylene is based on the decisive nature of his experiments and his priority in publishing and patent filing.

It should also be noted that such 'soft spots' as critics have cited in the organisation or program of Ziegler's laboratory (Ch. 5) have little if any bearing on the polypropylene question. More elegant kinetics or more sophisticated chemical engineering would not have accelerated polypropylene's debut; it was primarily a matter of relative emphasis in program targets. But there was also an uncharacteristic lack of persistence in following up an ambiguous experimental result, attributable, at least in part, to a predilection for a preconceived conclusion.

Following the initial discovery of polyethylene, there were two logical paths to pursue to find the boundaries of the new field that had been opened: (1) What other catalyst combinations would work as well or better, and (2) what other olefins could also be polymerised? Ethylene, with only two carbon atoms, is the simplest and smallest member of the olefin family; propylene (with three) is the next member in series, is available in huge quantities from oil and gas refining and hence is the obvious next candidate.

Ziegler's initial emphasis was on the first of these two questions, and his natural tendency was to cover all the ethylene bases before sending anyone up to bat on higher olefins*. Furthermore, he already knew from previous attempts to use propylene in the 'growth reaction' that it was less reactive than ethylene and gave products of lower molecular weight.

Nevertheless, the higher-olefin question was by no means entirely neglected. Breil was asked to investigate propylene also, and he tried it the very next day after his successful polymerisation of ethylene with zirconium as co-catalyst. The timing could scarcely be criticised, but the execution left something to be desired, and as a consequence his result was inconclusive.

Breil was accustomed to following the drop in ethylene pressure as an indicator of the amount of reaction, but with propylene, which is highly soluble in the solvent used, the pressure was low at the start and would be reduced only slightly by partial reaction. Breil apparently did not give this factor sufficient consideration, for when he saw that the pressure was not dropping noticeably, he added some ethylene in order to see if the catalyst was still active. He now saw the familiar pressure drop and, when he opened the reactor, found some polymer. He assumed this to be polyethylene, whereas actually it may well have been a mixture of polyethylene with at least some polypropylene. Ziegler said later that he believed a co-polymer had been formed, but it is not certain that significant co-polymerisation would occur under these conditions.

Unfortunately, Breil did not work up the product in an analytical way that would have revealed its composition. Instead, he wrote in his

---

*Another demand had just been placed on his already thinly deployed forces by the happy discovery of a direct synthesis for aluminium triethyl (from aluminium powder, ethylene and hydrogen). Because of its importance for chemical syntheses as well as polymerisation catalysis, this new process also demanded immediate experimental effort.

laboratory notebook, 'it turned out that propylene cannot be converted into a high-molecular-weight polypropylene.'

This entry provides a classic text for the patent attorney's favourite sermon—that to conclude that something 'cannot' be done on the basis of one or two tests is unwarranted, and to put such a conclusion in writing is distinctly unwise.

Here, then, were several successive lapses in an otherwise thorough and painstaking research program: a poorly designed experiment, insufficient examination of the product and a premature leap to a generalised negative conclusion. Ordinarily, none of this would have mattered, because the next experiments in the series would have quickly revealed the true situation. But in this instance there was a delay of several months before work with propylene was resumed. The delay was imposed by the decision to re-tool for 'greater efficiency'. The small, high-pressure reactors used up to this point were poorly suited for the new conditions of low pressure, low temperature and large volume (to accomodate solvent). Since the Institute had its own well equipped shops, they were commissioned to make a new set of vessels of more appropriate design. Pending their delivery, the work for which they were ordered, including that with propylene, was held in abeyance.

Actually, the need for better equipment did not compel the hiatus; had there been a sense of excited urgency, ways would have been found to make do with equipment at hand, even glass*. But it was easy to accept postponement in view of the negative interpretation of the first experiment, the evident predilection to consider higher olefins difficult if not impossible to polymerise and the preoccupation with ethylene.

In May 1954, the new reactors were ready, and shortly thereafter (June), Heinz Martin used them for a careful test of propylene's polymerisability. He quickly succeeded in making high polymer in good yield. By now, Ziegler had a patent attorney; he was given this new information, but neither he nor Ziegler assigned it urgent priority. In due course (August), he filed a patent application, naming Ziegler and Martin as co-inventors. Meanwhile, Ziegler, living up to his agreement with Natta, had sent a sample and description of the new polymer to Natta

---

*Professor Parkinson, of 'Parkinson's Law' fame, has pointed out that most research laboratories do their most creative work before they are 'properly' equipped. This appears to be another case supporting his point.

in Milan, only to discover that the Italians had already made it and had filed patents some ten days ahead of his!

Natta and polypropylene

In Natta's laboratory, the choice in emphasis between the two main lines of research—other catalysts or other olefins—was quite clear, and was just the opposite of Ziegler's. Ziegler had already been deeply involved with metal-organic compounds for years and was testing every combination in sight, but had concentrated his attention on ethylene as the monomer. Natta therefore turned to the question of higher olefins and gave that immediate priority.

Prof. Piero Pino, then a senior member of Natta's staff and now Professor at the Swiss Federal Institute of Technology at Zurich, says that he and Natta were originally more interested in synthetic rubbers than in plastics, and thus were more interested in propylene than ethylene because they believed it would polymerise to an elastic, rather than a plastic, material (they reckoned without the pronounced crystallisability of the stereoregular polymer). Later, they were to return to that objective with a modified, and successful, approach.

After receiving full information on Ziegler's polyethylene work, Natta discussed program ideas with Drs Pino and Mazzanti (another senior staff member), and suggested trying propylene immediately. They looked over the various catalyst systems disclosed in the Ziegler patent applications that Magri had brought back from Milan, and decided that the aluminium triethyl/titanium co-catalyst system offered the best prospects.

Paolo Chini, another of the 'Three Bright Boys', who had by now returned to Milan, was assigned to the task. On 11 March 1954, he made the first attempt. Starting in glass equipment (!) with no heat or pressure, he, like Breil, observed little reaction. He then transferred the reaction mixture to a pressure vessel; but instead of then adding ethylene, as Breil had done, Chini increased the temperature and the propylene pressure. This speeded up the reaction and produced a significant amount of polymer, which was worked up to a white powder (by removal of a sticky fraction by extraction with ether).

Chini immediately showed his product to Natta, who was in his office next door to the laboratory and thus readily accessible. Italians are traditionally emotional and voluble, but there is no record of any excited outbursts—only a laconic entry in Natta's own hand in his private notebook, an entry that rivals the British genius for restrained understatement:

'Today we made polypropylene.'

Presumably, Natta was not excited because he was not surprised. Polypropylene had been the object and the expected result of the experiment.

But even if the polymer's formation was not surprising, its form must have been. To Pino and Natta, who had visualised a rubbery product, the hardness of polypropylene was an unexpected phenomenon that called for an explanation. The answer was found quickly; the very next day a film pressed from Chini's first sample was stretched to five times its length and examined with x-rays. The diffraction pattern revealed a high degree of crystallinity, accounting for the rigidity and high melting point of the polymer. This prompt answer was a pay-off from the fact that Natta was familiar with, and understood, x-ray crystallography, and had a qualified expert (P. Corradini) prepared to take and interpret diffraction patterns.

Just as Ziegler's tremendous background in organic compounds of metals made possible the discovery of 'Ziegler catalysts' and linear polyethylene, so Natta's background and training, reflected in the work of his laboratory, showed to great advantage in this first encounter with polypropylene. Chini had persisted in his initial experiment until he got a positive and meaningful result. The product was immediately fractionated to separate it into manageable components. And the powerful tool of x-ray diffractrometry was immediately applied and correctly interpreted to reveal the secret of its unique and unexpected properties.

It was not long before Natta's efficiently organised staff had polymerised a number of other monomers in addition to propylene. Thus styrene ($C_6H_5CH{=}CH_2$), which normally gives a clear, glassy, low-melting polymer, was converted to a crystalline (and therefore opaque) polymer with a very high melting point (230°C).

Thus began a beautifully organised attack on the problem of the precise structure of what turned out to be the stereoregular polyolefins. It yielded an impressive series of elegant and high-calibre scientific papers, paralleled by the intensive engineering and commercial development inaugurated by Montecatini and picked up and carried further by most of the major polymer and chemical companies around the world. Thus also began the course of action that would bring Natta and Ziegler together in Stockholm ten years later to receive joint encomiums, but in the interim would doom their nascent scientific collaboration to stillbirth. Moreover, it would also condemn most subsequent licensing, patent and commercial developments to unending controversy and litigation.

8 The Falling Out

The joint research and development agreement made in 1953 between Ziegler, Natta and Montecatini had set the stage on which polyproylene made its debut. The continued attack by the merged forces of the two laboratories could have given the world a model of fruitful international cooperation that would impress scientists and politicians alike and shower benefits on industry and mankind. The unlikely pairing of a German 'coal research' and an Italian chemical research institute, backed by their respective sponsors, made it possible to move rapidly in whatever direction Ziegler's discoveries might take them—jointly opening new paths in science, with new technology pouring through those openings to fill gaps in human knowledge and supply human wants.

The distinguished professor at Mülheim was well equipped with a modern laboratory, a *carte blanche* contract and a competent staff ready to follow wherever his mind took them. His Italian friend, the highly respected professor heading the Milan Polytechnic Institute, enjoyed comparable prestige and facilities. Although he was more subject to influence by his sponsor than Ziegler, that very closeness gave Natta the advantage that the impressive resources of a major, diversified chemical company, his country's largest, could be counted on to support and exploit promptly any significant findings.

And yet, it did not come off—at least not at all in that potentially happy fashion. Many of the scientific and practical benefits that one might expect from such a cooperative effort did indeed materialise, but the cooperation itself was short-lived. In a dismayingly brief time, collaboration and friendship were replaced by disagreement and recrimination, and the two eminent scientists were no longer friends or colleagues—were, in fact, no longer speaking. And from that sudden and severe falling out ensued a heavy fallout of business and legal ramifications that has not fully settled after twenty years.

Why this almost sanguinary outcome from so sanguine a beginning? What sowed the seeds of this quick-sprouting discord? The question merits pursuit, not to satisfy morbid curiosity or to fix moral or legal responsibilities, but because it is central to our objective. Surely the answer will tell something about the conditions that influence creativity and invention, and something about the critical factors that surround and jeopardise any international cooperation in either science or business.

A complete, objective, answer is not easily come by. Personal accounts taken when events are fresh suffer distortion from the heat of the moment, and by the time emotions have faded, so have memories. Fortunately, recollections need not be our sole reliance, for detailed documentation is available from the legal depositions taken in connection with subsequent patent litigation.

The Men of Montecatini

It takes only two to make a quarrel, but in this case there was help from some of the people in Montecatini, and we need to know something about them to understand their role. As we have seen, Giustiniani saw Natta frequently enough that the latter could not fail to be conscious at all times of his sponsor's interests as a friend and as head of Montecatini. In addition, Giustiniani made sure that his legal staff and development department kept themselves informed on Natta's work and lost no time in following up any promising findings with patent applications and development work. The two men charged directly with those responsibilities took them so seriously and worked so closely with Natta that they had a great deal of influence on subsequent events.

Dr Ing. Bartolemeo Orsoni, as manager of Montecatini's Engineering Department, was responsible for chemical development activities, including patents and licensing. Trained as an electrical engineer, he had develcped a very strong ambition to advance in management, and pursued it with diligence. The image he usually presented was that of a quiet, unemotional man whose primary objective was getting at the facts in a calm, objective manner. Within the company, however, he displayed strong competitiveness towards people who were nominally his colleagues*. His boss, Giustiniani, was reportedly impressed with Orsoni's vision of the things that could be accomplished by aggressive action, and may thus have been encouraged to lead Montecatini into undertakings that were overtly bold and not always successful.

Orsoni's mastery of both technical and commercial details gave him a degree of self-confidence that sometimes showed through his normally disarming exterior. An observer recalls his offering a top U.S. executive advice as to how the U.S. company should run their business—advice that was not fully appreciated. His strong and forcefully expressed opinions also enlivened the operation of joint companies that Montecatini formed in later years to exploit their polymer and patent positions in various countries.

Capitalising on the literal meaning of his name (*orsa* = bear; *orson* = big bear; *orsoni* = big bears), Orsoni is reported to have had four bear flags made and mounted on the four corners of his World War II jeep, an action that must have made him something of a local celebrity.

Dr Giuseppe DaVarda, Manager of Montecatini's Patent Department, was responsible for protecting Natta's inventions with patents assigned to Montecatini, and for negotiating with potential licensees. He was a determined, dedicated difficult man—an indefatigable negotiator who wore others down by persistence and reluctance to make concessions. Orsoni was his boss, but that was not always apparent when they teamed up in negotiating sessions, for Orsoni usually chose to provide a quieter obligato to DaVarda's sometimes strident lead.

But DaVarda was not without his graces or his champions. Even negotiators who found him intransigent had a grudging admiration for his dedication and acknowledged that he could be pleasant enough on social occasions. And at least one leading U.S. executive who engaged in many lengthy negotiations with Montecatini insists that he always found DaVarda forceful and persistent but not unreasonable, and rude only to those he considered amateurs or incompetents (a fairly large group, evidently). When he was not officially representing his company, he could be good company.

These two men are significant factors in our story because they were

---

*For example, Giulio Ballabio, who was responsible for some of the development work based on 'Natta chemistry' and Giorgio Mazzanti, who later became the 'number two man' in Montecatini but resigned in 1973 after the second or third drastic reorganisation.

involved so closely with Natta and played such key roles in shaping and interpreting the contract he made with Karl Ziegler. The very fact of their intimate involvement vitiated any possibility for another simple 'two-man' agreement (like the one negotiated between Ziegler and Borrows of PCL), which Ziegler would doubtless have preferred.

Portents of Problems

Signs of potential trouble in the German–Italian scientific entente developed quite early but were not fully recognised. The arrangement whereby Natta's 'Three Bright Boys' were allowed in Ziegler's scientific home worked smoothly only until it was put under strain. The strain came late in 1953 with the discovery of linear polyethylene.

The three Montecatini representatives had returned to Milan in August, but Magri was back in Mülheim in the autumn and learned of the new discovery very quickly through the natural medium of the excited conversations that ran around the laboratory. He promptly passed all he heard back to Milan in correspondence. Magri then spoke with Ziegler, told him that he had heard that a new high polymer of ethylene had been made, and asked for details. In Magri's words, Ziegler was 'very formal'. He confirmed the discovery, but said that it was not yet clear that it fell within the scope of his agreement with Montecatini. Therefore, he asked that the Montecatini representatives not press enquiries until that issue had been resolved.

Magri could easily agree, since he already knew the key details. Even after his talk with Ziegler, he passed additional information to Natta through Montecatini* and to his friend Chini in personal correspondence.

The degree of dissembling thus revealed will not surprise anyone familiar with common practice in 'industrial intelligence', especially at the international level. Nevertheless, had Ziegler been aware of it, he would surely have been offended and might either have broken off the agreement or at least asked the Montecatini men to leave. But it is likely

---

*Magri's letter of 15 November 1953, to Montecatini's Ballabio, read in part: 'In the meantime from indiscretions which I was able to gather before my talk with Prof. Ziegler, I know that one of the samples of polyethylene I got to see was coming from a test conducted with aluminium triethyl in the presence of zirconium at about 100 atm and a temperature of 100–110°C.'

that his disillusionment would have been less with what the brash young men did than with what his distinguished colleague in Milan did not do—i.e. extend to Ziegler the courtesy of letting him know what information he had obtained.

Regardless of what might have happened, nothing did, and the issue was defused by Ziegler's subsequent decision that polyethylene would come within the terms of the agreement with Montecatini (perhaps Ted Borrows had unwittingly done the Italians a favour in establishing a precedent when he insisted on such an interpretation of PCL's contract). After Ziegler came to that conclusion, he gave Magri full information and authorised its transmittal to Natta. But that act, in turn, lit the fuse on a much greater bombshell.

Denouement and Disillusion

Ziegler visited Milan in March 1954, to complete his formal disclosure of the details of the polyethylene discovery to Natta and Montecatini. It was characteristic of him that he did not delegate this activity to a deputy or rely on written communication. It is also characteristic that he scheduled the visit as a stop on his way to a vacation in Sicily.

The meeting was attended by Natta, Orsoni and DaVarda. After completing his description of the new discovery, Ziegler made one important comment. He freely acknowledged Montecatini's right to full and up-to-date information and to whatever development and commercial exploitation they chose to make, and indicated that he expected to continue the exploratory research and the widening of the newly opened field of science.

'Let us for some time alone in the field', he asked.

This was agreed to on the spot; in fact, the Montecatini representatives proposed to make a written record of the understanding. So a 'minute of the meeting' memorandum was drawn up which spelled out Ziegler's commitment to exploring new ways of making aluminium alkyls and catalysts for the polymerisation of olefins, especially ethylene.' In the light of subsequent events, the fact that the memorandum refers to olefins as a class, and not solely to ethylene, was later stressed by Ziegler as 'very important' (Montecatini was to contend, however, that the reference was only to propylene dimers and trimers). The same document noted Natta's intent to investigate certain aspects of the field, including reaction mechanisms and x-ray examination of catalysts and polymers.

The memorandum was signed or initialled by the parties on 9 March and Ziegler went off to his vacation in Sicily and thence back to Mülheim, confident that he had fulfilled his obligations and under the impression that the agreement gave him a clear field to continue his research.

Two days later, Chini had made his first propylene polymerisation experiment in Natta's laboratory and Natta had recorded its success in his own diary!

At this juncture, one might visualise Natta calling or wiring his German colleague to share the news of the discovery and arrange for sending him samples and for the division of effort for further research. Certainly the positive result was exciting enough to sweep away any question as to how Natta came to be doing the kind of research that Ziegler was planning to do himself. But scientists display very human traits, including ambition, and corporations are dominated by commercial considerations. What actually happened was that Montecatini's lawyers filed a patent application on polypropylene with Natta as inventor.

Two months later (19 May 1954), Natta and Orsini returned Ziegler's visit. According to his later testimony, Natta took advantage of that occasion to ask Ziegler whether he had polymerised propylene. Ziegler said he had not, but that he had co-polymerised it with ethylene. Natta then asked him 'how the situation would be' in case propylene were found to respond to Ziegler catalysts. Ziegler replied that the question need not be considered, since propylene would not respond. Natta pressed the question.

Ziegler finally dismissed it by saying that he had tried propylene alone and that it did not work: in the phrase now famous from its quotation in ensuing litigation and made classic by its use in warnings patent attorneys give their clients: 'Es geht nicht!' Further, Ziegler implied, if he had not been able to polymerise propylene, no-one else need waste his time trying.

Natta let the matter rest there. Most significantly, he did not tell Ziegler that he had already made solid, crystalline polypropylene. Natta explained subsequently that 'I did not tell him because I had to take patents first. I asked whether he had polymerised propylene in order to know whether my process was new'.

Natta's actions in this matter have puzzled and pained most of the people who are familiar with the circumstances and the persons involved. Whether they are characterised as questionable ethics or simply good sharp competitive practice depends on who is doing the characterising, and in what country.

The basis for the misunderstanding seems clear: from Natta's viewpoint, the arrangement with Ziegler was not a symmetrical, cooperative research project in which both parties were committed to full exchange of information in a two-way flow. After all, the money flow had been all one way; Montecatini had paid Ziegler 600 000 DM for the information and license to use it. No doubt they and Natta considered themselves free to make any improvements or extensions of the technology that Natta's brains and Montecatini's money could accomplish. What is difficult to understand is that they did not make all this clear to Ziegler and why they permitted his misconstruction of their intent by agreeing to his proposal for division of research efforts, only to act in quite contrary fashion as soon as he was out of sight.

To Karl Ziegler, therefore, the revelation came as a rude shock. True, he had already had occasion to be unpleasantly surprised ('a little astonished', he put it) at one action of his Italian friend, when Natta published certain measurements on 'Ziegler polyethylene' which at the time were still confidential. But that did not deter him from acting as he felt was expected of him under his agreement with Natta and Montecatini. As soon as Heinz Martin had succeeded in making a respectable propylene polymer, Ziegler had sent a sample off to Milan, with a covering letter to Orsoni which ended with the statement, 'finally reaching the conclusion that our most active polyethylene catalyst also gives a beautiful polypropylene.'

The replies he got from Orsoni and Natta were not at all what he expected. Orsoni's stated that Natta had already made polypropylene and Natta's described some of the properties 'of the polypropylene we have made ourselves'. Ziegler thus discovered that Natta had begun polymerisation research immediately after, and contrary to, Ziegler's understanding of their agreement, and that he had already succeeded with propylene when he questioned Ziegler about it in May.

Ziegler has described his reaction in what is undoubtedly a careful understatement:

'I was a little bit shocked by what I learned from the two letters of Drs Orsoni and Natta, and if you are a little bit shocked you shall not write letters'. So he sent a 'relatively short' (curt?) reply and went to Milan a few weeks later to express his opinion and state his position in strong words.

Although he maintained a dignified restraint in discussing these events with others, there is no doubt that he was not only shocked but humiliated and angered as well. Humiliated, because these rude surprises arising from a situation much of his own making made his judgement look bad (his reliance on understandings between individuals now looked naive and old-fashioned) and because his 'do-it-yourself' patent writing had not given him the broad coverage that could have protected him regardless of what Natta might do. Angered, because he now stood to lose the recognition and priority of discovery that otherwise would have been his, and to miss out on some money if polypropylene proved to be a winner (how big a winner and how much money, neither he nor Natta could have guessed at the time).

The high hopes for continued fruitful international collaboration between the two prestigious institutes now all came crashing down; in fact, Ziegler and Natta for a time were not even on speaking terms. Ziegler reluctantly brought his attorney (von Kreisler) into the picture and began the hard job of salvaging as much of the situation as possible via a painfully hammered out legal agreement dividing up the licensing rights. This required extensive negotiations with Montecatini, in the person of Dr DaVarda, which, as Ziegler remarked, 'were not always a very agreeable thing.'

But some sort of working arrangement was urgently needed by both parties, and so a division of the field was finally arrived at that enabled each to deal with prospective licensees on a consistent basis. In brief, it provided that Ziegler would license the basic catalysts and the polyethylene process, Montecatini the polypropylene process. Royalties collected on the later process would be divided, 30% to Ziegler and 70% to Montecatini.

Putting the best face on the matter, Ziegler wrote to Montecatini (August 1955), confirming the arrangements concluded by the respective legal representatives whereby 'our previous considerable dissents as to the interpretation of our first agreement and the notion of common invention have been eliminated' and saying that he was 'looking forward to a further pleasant cooperation between your company and myself.' Natta was not mentioned.

Thus, out of necessity, the mother of agreements as well as inventions, the rift was patched over, although it was never fully healed. It is one of the surprising aspects of the affair that it never really got into either the scientific literature or the popular press. Disagreements between scientists over priorities or one's use of another's results are often recorded in vigorous polemics that enliven, if they do not enrich, the literature of science. It appears to Ziegler's credit that he did not choose to air his grievances in this form. In fact, he and Natta re-established a stiff politeness toward one another* and took pains (usually) to acknowledge each other's contributions in their subsequent publications.

In fact, it was Natta, not Ziegler, who christened 'Ziegler Chemistry' in one of his earliest publications. The first papers (December 1954) which announced the discovery of crystalline polypropylene and other polyolefins and characterised them as having an alternating DLDLDLDL ('isotactic') structure, revealed nothing as to how they were made, saying only that such information would be published later. The 'cooperation of the Montecatini Company' and the contributions of some dozen coworkers were acknowledged, but no mention was made of Karl Ziegler. But a half-year later, in a lecture given in Zurich, Natta stated: 'We have prepared these polymers for the first time by polymerisation of propylene with the aid of catalysts of the type discovered and applied by Ziegler in his brilliant polymerisation process of ethylene at low pressure', and thereafter he referred repeatedly to 'Ziegler-type catalysts' and 'Ziegler polymerisation'.

Similar careful courtesies marked their references to each other in discussions with third parties, including prospective licensees. In this, they may well have been following advice of counsel – to do nothing to aggravate an already sticky situation in which a good deal might be at stake. It is unlikely, however, that counsel or chemist foresaw the full ramifications or the size of the ultimate stake.

Why?

Natta could not have been unaware that his views of his working arrangement with Ziegler and its mutual obligations were quite different from Ziegler's, and must therefore have known that the latter would be upset, perhaps antagonised, when he learned what Natta had done. The question then is why, knowing the probable consequences, did he do as he did? He had already 'arrived' in his scientific career, his reputation was already established and his position as director of the Polytechnic Institute secure. The field that was open to him via the Ziegler agreement had unexpectedly developed exciting new prospects that could be exploited

*In a poignantly revealing slip, Ziegler once remarked in discussing an event that preceded the falling-out, 'at that time my relationships with Dr Natta were quite friendly'. The he hastily added: 'they are still friendly.'

effectively through cooperative research, and promised glory and money enough for all. So he had no need to try to squeeze every possible advantage out of the situation at the expense of his German friend.

Moreover, Natta has spoken with pride of the 'moral rigour' (*rigore morale*) which characterised his research. Why, then, did he pursue a course of action that was apt to alienate his collaborator and terminate their collaboration?

There is no lack of proposed answers to this question; the difficulty is to identify the one closest to the truth. We are reminded that scientists are human and that the pursuit of human ambition is often a greater driving force than the esoteric love of science for its own sake. By personal account of these and other Nobel Prize-winners, particularly the remarkable self-revelations in Watson's '*The Double Helix*'*, we know that even that prestigious award is sometimes stalked by scientific big game hunters with cool and savage intent.

There is no doubt that Natta was ambitious, but ambition itself is not necessarily a failing, as was so ably stated on behalf of another, earlier, famous Italian (Roman) by Mark Antony. It simply does not stand to reason, however, that a man like Natta would let his personal ambition override considerations of friendship and professional ethics, nor is it likely that anyone could have been thinking about Nobel Prizes at that early stage.

Accordingly, one is inclined to look beyond Natta himself. One observer has suggested that Natta, the true scientist interested primarily in science, was under the influence of Giustiniani, who, as head of a commercial enterprise, had to be concerned primarily with commercial aspects and would naturally favour actions that would be to Montecatini's benefit.

Another opinion would attribute the primary influence, not to Giustiniani, but to DaVarda. Giustiniani, after all, was the head of a huge, diversified chemical business, and could not have had either the time or the detailed technical knowledge to intervene first-hand in this sophisticated research field. DaVarda, on the other hand, had that subject as his responsibility, and he was equally devoted to furtherance of Montecatini's interests. As one observer commented: 'His dedication to the concept of Montecatini as the greatest chemical company in the

---

*Watson, *The Double Helix*, Atheneum, New York (1968), c.f. also the Guillenin–Schally controversy (*Science*, April 22–May 5, 1978).

world led him to fight for it even at times when fighting was unnecessary'. Some of those who observed at first hand his single-minded approach to negotiations would be surprised if he did not always put Montecatini's interests first.

But others who knew these men better say that that, too, is a misreading of character. The voice, they say, was the voice of DaVarda, but the hand was the hand of Orsoni. The latter was the boss, and his influence, whether exerted directly or through DaVarda, must have been the dominant one.

Still other names get mention as possible contributors to the deterioration of the Ziegler–Natta relationship. However, it would be a mistake to cast Natta in a naïve or passive role as one deferring to others in all matters outside the laboratory. As we have seen, he was no ivory-tower scientist but a thoroughly practical man, sure of his goals and proud of his understanding of industrial goals.

An interpretation somewhat different from all these is probably nearer the truth. Invoking historical perspective, the hypothesis may be advanced that no individual formulated and promulgated any deliberate plan to take maximum advantage of the situation; nor was there any group that got together to form such a plan—for the reason that no such decision and no such special meeting was necessary.

Italy was then engaged, for the second time in this century, in a determined effort to break out of the 'POP' (painting, opera, pasta) cultural tradition and take its place among the industrialised nations of the West. The Mussolini regime, whose vaunted ambition was to accomplish this great leap in a few years, had ended with the country still worse off, an outcast and orphan among the industrialised victors and her re-industrialising partner in defeat.

Italy's technical and industrial leaders were determined not to lose out all over again in the post-war economic race, and to that end they bent every possible effort and every available means. All other considerations, even personal fame, and certainly the luxury of lofty intellectual and ethical standards, were secondary and subservient to that goal. This spirit so imbued the key men in business, commerce and even science that all significant actions were guided by it without the necessity for long deliberations or consultations.

This pervasive attitude is evident in the phraseology of the press accounts of Natta's Nobel Prize award:

'Natta is not only a scientific genius, but a man who has demonstrated how the investment of money in scientific research is the best if not the only future solution for countries poor in prime resources such as ours.'

'But at this point we must remember another factor: that of international prestige.'

'Our country has demonstrated in various sectors of modern science to be able to contribute first order brains.'

'They [young scientists] represent the most precious investment that a country such as ours can have. It is they in this field [who are] our primary natural resource. It is they who assure a future for our industry, to our [trade] balances, to our active presence in the international industrial field.'

If all this sounds like a large measure of narrow patriotism to ascribe to any people in modern times, we may note that what served the country's goals, as they were conceived in this context, also conveniently served some personal and corporate goals. The loose cloak of patriotism is often flung round the meaner garments worn closer to the skin.

As with all assignments of motive, this interpretation is subjective, but it is in keeping with the factual record of the actions of the Montecatini Company over a period of years. Those who worked with, or competed with, Montecatini in those days have remarked that the company's concept of corporate ethics at the time was rather remarkable, even in the genteel jungle of international business.

The records of patent prosecutions and interference proceedings is illuminating. An internal memorandum of Montecatini's which refers to 'some rash and pessimistic statements by Dr Ziegler' suggests intent to take advantage of the man and the situation. Montecatini representatives complained that the U.S. patent system penalised them as foreign applicants* and unreasonably delayed granting their patents. Yet a memorandum of instruction to their patent attorney in the U.S. reads, in part:

'arrange things so as to involve all our competitors in an interference limited to the crude polymer. On any occasion in which this would prove convenient one should not hesitate to select a long and destructive procedure. Try thus to avoid or at least delay as much as

*As it does. A foreign patent applicant can claim only his filing date as his earliest priority, whereas a U.S. applicant, presumably because his records can more readily be validated (but perhaps also deliberately to give him an advantage?) can 'swear back' to notebook records of conception and reduction to practice.

possible allowance of claims to our competitors.'

(Of this statement, one of the opposing attorneys commented, 'I think that, if this case is ever put to rest, would be a suitable epitaph!').

Due in part to such manoeuvring and in part to dogged persistence of the U.S. companies opposing Montecatini*, the proceedings in the U.S. patent courts have been among the longest (over fifteen years), most complex (several different actions, half-a-dozen defendants, tens of thousands of pages of testimony), and most acrimonious on record.

The major litigations, at long last, have now been settled, in or out of court. But the transcripts of those long-drawn-out proceedings, including particularly the testimony of the principals, remain as a permanent, detailed and public record of what transpired. They contain many illuminating passages that provide insight, not only into the legal issues, but also into the actions, attitudes, and, to some extent, the motives, of the individuals and organisations involved. An attorney representing Ziegler in patent interference proceedings characterised them as 'an extremely hard, and I must say, bitter, proceeding in the Patent Office with Montecatini in which I do not consider the dealings among counsel to be of the friendliest nature or on the highest point.' Later, he remarked: 'I consider the conduct of the whole interference outrageous', and castigated the 'politics and dirty politics' involved in it. Even with due allowances for the hyperbole of advocacy, these are strong indictments.

Whatever the root and, possibly, multiple, causes of the falling-out between Ziegler and Natta, that unfortunate event provides a classic example of the morasses of misunderstanding that can easily be encountered when complicated and intimate arrangements are made, even with the best of intentions, between parties of different nationalities, languages, traditions and aspirations. In retrospect, Ziegler might well have said, 'All this shows that nothing takes the place of a "gentlemen's agreement" between people who understand and respect each other'. On the same evidence, a lawyer might say, 'It shows that nothing takes the place of a professionally drawn documented agreement'. The general history of agreements shows that, in fact, both can work, that

---

*'The trouble with you Americans is that you never know when to quit or how to gracefully acknowledge defeat', was the complaint one former Montecatini employee made to me. He was tired of being summoned repeatedly to testify at hearings in Europe and the U.S. John Paul Jones would have taken his comment as a compliment, but none was intended.

both are vulnerable to the impact of unforeseen eventualities, but that, of the two types, the 'personal understanding' is the more fragile.

PCL and Polypropylene

There is an interesting contrast between the attitudes and actions just described and those of other groups in other countries. In England, Petrochemicals Ltd were attempting to cash in on and live up to their agreement with Ziegler by operating a sizeable pilot plant making polyethylene with Ziegler catalysts. One day the technologist in charge of the plant (Bernard Wright) telephoned Ted Borrows the technical director, to tell him that the pilot plant run planned for that day was all set to go except that the ethylene supply had been cut off by a line failure.

'Do you mind if I try propylene instead?', he asked.

'Sure, go ahead and try it', said Borrows. 'It's bound to work, I should think.'

A few hours later, Borrows' telephone rang again.

'It's working, it's working!', cried Wright. 'Come and see!'

Borrows by now had a VIP visitor, the head of the company, in his office, but they both went out to the plant area and saw polypropylene coming out of the reactor. They were delighted but not terribly surprised; Borrows had not heard Ziegler say, 'Es geht nicht!', and knew of no theory or evidence that said propylene would not polymerise.

With this fine result in hand, Borrows might have fired off a patent memo or a cable to Mülheim, or both. Instead, he did neither. He never even considered filing a patent or attempting to establish priority of publication, for such actions would not be in keeping with the terms of the licensing agreement or the 'gentlemen's understanding' he had been at such pains to reach with Ziegler.

But neither did he bother to notify Ziegler. To him, trying propylene was obvious after ethylene, and he assumed that the much larger and more sophisticated research group at Mülheim had either made or would make polypropylene also. He further assumed that Ziegler's patents would cover propylene as well as ethylene and that rights would flow to PCL automatically, just as they had on polyethylene.

It would have been an act of prudence to have checked with Ziegler to make sure that all these assumptions were sound, but it would not have altered the course of history. Ziegler had, in fact, made polypropylene by that time, but he and his patent attorney had not acted with the urgency that would have been necessary to beat Natta to the patent office.

Hoechst Hustles, Holds Back

Ziegler had, perhaps, relied overmuch on acceptance in countries outside Germany of the German scientific tradition that when a scientist opened a new field of research he should be allowed a reasonable interval in which to broaden his investigation of that field before others moved in. Inside Germany, that tradition had a better chance of being understood and honoured.

Hoechst, one of the German chemical companies entitled to information and a licensing option under Ziegler's contract with the Max Planck Institute (and to which he had been introduced as a prospective employee while still a university student (Ch. 3)), had became one of Ziegler's first licensees and an early producer of 'Ziegler polyethylene'. As did most licensees, Hoechst lost little time in starting some research on 'Ziegler Chemistry' in their own laboratories. According to the testimony of members of the 'Polyethylene Committee' that was set up to coordinate that research, the committee decided that polymerisation of propylene should be attempted also. In a short time, the committee received a report that one of Hoechst's research chemists, Dr Rehn, had succeeded in making polypropylene, using a 'Ziegler catalyst', in March 1954.

However, after due deliberation, the committee 'decided for the time being not to continue any research on polypropylene and not to file a patent application in order not to usurp Ziegler's research field.'*

*One observer of this scene remarked that, in this case: 'Honour is without profit, save in one's own country.'

9 Harvesting the Fruits of Innovation — Polypropylene

> It is perhaps the first time in the history of macromolecular chemistry that a scientific discovery has been followed so rapidly by such a vast amount of research in scientific and industrial laboratories.—Giulio Natta

The scientific community and the general public had heard nothing of the Ziegler/Natta collaboration or contention, but they suddenly became very much aware of Giulio Natta when his discovery of polypropylene made the news. As in the case of Ziegler polyethylene, the word spread fast and far, and through many of the same channels. Once again, Herman Mark was among the first to know. He visited Natta soon after the latter had written, 'Today we made polypropylene', but before he had published, or Montecatini had announced, the discovery. Natta told Mark that he had a crystalline polypropylene, but pledged him to secrecy. This must have put Mark in a somewhat uncomfortable position for a time, since the analogy to Ziegler's crystalline polyethylene was obvious, yet he was not free to discuss it with his German friend.

As soon as the restriction was lifted and Mark could play his usual role as a scientific circuit rider, he spread the gospel effectively. When, one day early in 1955, I visited him at Brooklyn Polytechnic Institute, he seated me close to his crowded desk in his tiny office and opened the conversation by asking:

'Well, have you seen the new polypropylene and polystyrene?'

When I shook my head, he handed me a sample.

'Here is crystalline polypropylene, melting at 150°C.'

Then he gave me another chip, about the size of his thumbnail.

'And this is crystalline polystyrene. It melts at 250°C.'

To one who had struggled with the limitations of the known polymers, it would have been no more startling had he said: 'Here is an ice cube that won't melt below 60°C'. I was quite familiar with propylene 'polymers' that were oily liquids, and had personally tried to convert them to useful solids by massive crosslinking reactions, with no more success than to get a 'cheesy' gum. Yet here was a solid, translucent polypropylene which was not only high-melting but microcrystalline. The only known polystyrene (of which Mark was one of the inventors), a clear, amorphous glass melting around 100°C, was an equally far cry from the crystalline polystyrene he was showing me.

Mark could give me no details at that time except that the polymers were made under simple and mild conditions. I knew he was too sharp to have been taken in by any flummery, but I had no way to judge the practicality of whatever process had made those samples. I convened a special meeting at the Shell Development Company research laboratories to pass on this startling information to interested but incredulous chemists, but I could offer no corroborative detail, no explanation and no evaluation.

Similar discussions must have been occupying puzzled research groups in many different laboratories. A few companies with direct sources of first-hand information had more to go on. Thus, Petrochemicals Ltd learned about Natta's polypropylene, not from Natta or Mark, but from Karl Ziegler. But it was not an exciting surprise to them, because they had by that time already made polypropylene themselves (Ch. 8).

In the U.S., the duPont Company doubtless also heard about Natta's discoveries from their consultants. These included Paul Flory, whom we met earlier (Ch. 2), but he was apparently not first with the news. Flory attended a meeting in Turin in the autumn of 1954 at which Natta gave a paper on a chemical subject. When they met, Natta apologised for not presenting anything very new or exciting in his paper.

'I have something new', he said 'but I cannot talk about it yet. Come back in the spring and I can tell you.'

Flory attached no great significance to that remark but remembered it when, in January 1955, he received a letter from Natta enclosing the manuscript of a 'Communication to the Editor' which Natta had submitted to the Journal of the American Chemical Society, announcing the discovery of stereoregular polypropylene. It had been rejected by that journal on grounds that their policy required that any discussion of a new product include a description of how it was made. Natta explained in his letter that he was not yet free to disclose such information but was still anxious to have the discovery published. Flory could read between the lines an appeal: could he, Flory, who was on the editorial board, persuade the journal to reverse its decision?

Flory did convince the editor to make an exception in this case, and the 'Communication' was published in the spring of 1955, constituting the first announcement of the discovery in an American journal*.

Natta's letter also repeated his invitation to visit Milan, and Flory did so in April 1955. Natta received him very graciously and told Flory that he was indebted to him for the clue to the structure of polypropylene. Just after writing in his notebook, 'Today we made polypropylene', Natta had spent a weekend at his mountain chalet in the Italian Alps and was 'relaxing' there by reading Flory's then-new book, *'Principles of Polymer Chemistry'*. The discussion therein of the possibilities for crystallinity arising from steric order in polymers and the reference to Schildknecht's crystalline polyvinyl ethers suggested to Natta that this might be the explanation for the surprising crystallinity of the newly discovered solid polypropylene. Back in the laboratory, he was quickly able to confirm the supposition.

Flory's hospitable reception in Milan was thus understandable. He was even introduced to Giustiniani, who encouraged 'showing him everything'. Although disarmed by the academic appearance of Natta's Institute, housed as it was in the University of Milan, Flory took the precaution of mentioning that he was a consultant for duPont, but was assured that this posed no problem. Giustiniani, with his penchant for throwing a figurative arm around the shoulders of the mighty, explained that: 'We are on very good terms with duPont'.

Flory therefore felt free to advise duPont that he had learned of some interesting developments on stereopolymers in Natta's laboratory. He was puzzled when duPont showed no interest in hearing more and, in fact, put him off on two separate occasions. Later, he learned the reason: duPont had already made solid polypropylene in their own

---

*Natta, G., *J. Am. Chem. Soc.*, 77, 1708 (1955).

laboratories* and did not want to become 'contaminated' with outside information†.

Hercules pioneers again

As we noted in Ch. 6, Hercules had finally geared up for action, after first dropping their option for an exclusive Ziegler licence and then coming back for non-exclusive rights. When the team that went to Mülheim to pick up all the known technology returned to Wilmington, Hercules not only began plant design work but also launched their own research program on Ziegler chemistry. Within a few weeks, E. H. Vandenberg, at the Hercules Research Center, began studying polymerisation with Ziegler catalysts.

Vandenberg had been trained as an engineer at Stevens Institute of Technology, but when he joined Hercules in 1939 it was as a research chemist. Having escaped exposure to Ziegler's dictum that other olefins 'could not' be polymerised, he was almost as quick to extend his testing to propylene as Natta had been. After only a week's work with ethylene, he made his first attempt to polymerise propylene.

Vandenberg had hopes that propylene would give a high polymer with Ziegler's catalyst, but he had no firm preconceptions as to what kind of a polymer it might be. He would not have been surprised had it turned out to be a rubber, and he certainly did not expect to get a crystalline, high-melting plastic. His first experiment gave only 92 mg of crystalline polymer, but it was enough to discover that it had a high melting point and that it could be pressed into a film and drawn into a fibre.

To Vandenberg, the surprising properties of his first polypropylene sample showed that it had the potential of being a useful plastic. This was an exciting lead, but his pursuit of it was delayed by a previously scheduled three-week vacation. Immediately upon his return, larger samples were prepared, more extensive evaluations were made and diligent efforts were begun to improve polymer yields. Within a few

---

*DuPont patent disclosures indicate that they had made poly-α-olefins from which crystalline fractions could be separated.

†Giustiniani's sweeping assurances did not prevent Montecatini's lawyers from raising the question later as to whether duPont had been aided by confidential information obtained through Flory.

months, sufficient progress had been made that a commercial process appeared feasible.

Neither did it take Vandenberg long to recognise the probable source of polypropylene's remarkable properties. The American Chemical Society meeting at which Schildknecht first described his vinyl ether polymers and speculated on their possible stereoisomerism (Ch. 2) was attended by Vandenberg, who also gave a paper there. He heard Schildknecht's presentation and was sufficiently impressed by it that he could readily visualise, when polypropylene appeared on the scene, that it might be another example of such isomerism.

Meanwhile, both Ziegler and Natta had made polypropylene, and word of this came to Hercules, again through the personal contacts of Arthur Glasebrook. Some years earlier, at the 75th Anniversary meeting of the American Chemical Society in New York, one of the special features of the meeting had been the attendance of a number of foreign chemists whose travel expenses were underwritten by the Ford Foundation. They were also invited to visit some American laboratories, including Hercules' at Wilmington, and it was there that Glasebrook, who was then working in high-pressure chemistry, met Piero Pino from Natta's laboratory. Thus began a friendship that proved beneficial later.

While visiting Ziegler's laboratory in the summer of 1953, Glasebrook was invited by Pino to Milan, where they talked about Pino's research (at that time on reactions of carbon monoxide) and Pino introduced Glasebrook to Giulio Natta. In the autumn of 1954, both Glasebrook and Pino attended a meeting of the Oil and Chemical Society in Essen, Germany, and on that occasion Pino gave his friend some inkling of the polypropylene discovery. The next summer, Glasebrook again visited Milan and was shown fibres from polypropylene (also from isotactic polystyrene, which at that time they thought might be equally interesting).

Even though they missed out on patent priorities and Nobel Prizes for Vandenberg's independent discovery of polypropylene, Hercules' research yielded a number of useful and patentable process improvements, and this time the opportunity to exploit their position was not lost. It must have been galling to find not only that they had missed the polypropylene patent by a few months, but that their broad Ziegler licence, which covered all olefins, probably would not spare them the necessity of taking a licence from Montecatini, since Ziegler's own patent coverage was too narrow. Nevertheless, they bit that bullet, took a licence from Montecatini, and meanwhile moved quickly enough to repeat their pioneer role and become the U.S. leaders in polypropylene.

Again, there was much opposition within the company. Some informed members of management were struck by the striking technical shortcomings of the embryonic product; others thought that Hercules had swallowed too big a lump for even their giant* to digest when they took on Ziegler polyethylene, and that rumination should precede diversification.

But Paul Johnstone, who had been unenthusiastic about polyethylene, became a vocal and indefatigable advocate of the polypropylene project. His persistence ('My colleagues call it bullheadedness', he told me) was an important factor; another was the possibility of taking advantage of the two-line polyethylene plant just coming on-stream and converting one line to polypropylene production with minimum time, risk and cost. The project was approved in 1958 and production was achieved in the same year.

The operation of the plant and the quality of the product left much to be desired, but so did the receptivity of the market. Fortunately, both proved amenable to improvement at about the same pace. Johnstone had seen to it that the propylene feed came from a different supplier than the ethylene, and no raw-material problem delayed this project. Within twelve months from start-up of this first U.S. plant, Hercules were selling all they could produce, and a much larger plant was on the drawing boards.

In rapid succession, management was asked for and approved additional capital to build facilities for still more polymer production, for polypropylene fibres and for polypropylene film. These investments gave Hercules a running start in the U.S. and a lead that it has never relinquished.

Some years later (1969), the Commercial Chemical Development Association, an organisation of product development specialists in the chemical industry, awarded its annual medal to Elmer Hinner, Hercules' board chairman, for the successful introduction of polypropylene to the American market, while Hinner was Manager of the Polymers

---

*The Hercules corporate logo depicts the famous giant swinging a big stick.

Department. In his acceptance speech at the award banquet, Hinner acknowledged the contributions of others in the Hercules organisation and invited them to stand and be recognised. Among those who responded and those who watched must have been some who saw a certain irony in the blanket acclaim of a group that included the chief contributors to solving the problem and others who had been part of it.

Standard gets into the act and into high gear

When, in the spring of 1955, Natta announced his discovery of polypropylene, there was at least one individual in the laboratories of Standard Oil Co. (N.J.) who was not surprised, although he was probably greatly disappointed.

Dr S. B. Lippincott, who, as we saw in Ch. 6, had begun the study of Ziegler-type catalysts, was a proficient 'chemist's chemist', an organic chemist from Purdue who came to Standard after six years at Commercial Solvents Corporation. His earliest experiments on Ziegler chemistry led to still another instance of separate, but parallel, discoveries. For he, too, tried Ziegler catalysts on propylene. By February of 1955 he had made a polymer of propylene from which he could separate a solid fraction. He was impressed by its high melting point and recognised what that could mean in terms of practical applications*. Unfortunately, in one of the abrupt shifts that characterise corporate staffing policies, Lippincott was transferred to another project only two months later. Thus, momentum and opportunity were lost, both to Standard and to Lippincott. He could soon see that his being sidetracked had caused him to miss a very large bus, and his disappointment and disillusionment endured throughout the rest of his career.

Overall, however, Standard could not be faulted for lack of diligence in pursuing this new field of opportunity. The pilgrimage to Milan that followed Natta's polypropylene announcement was joined by William Asbury, who hoped to add a licence from Montecatini, if needed, to the one he had previously negotiated with Ziegler. However,

---

*Another chemist, Larry Eby, also made polypropylene about the same time, using a 'bomb' (high-pressure reactor). However, Lippincott's experiments had added significance because he carried out the polymerisation at low pressure and fractionated the polymer to isolate the solid portion.

he was not met with what he considered a good reception nor his enquiries with proper consideration (he was not the only would-be licensee who had that reaction).

Annoyed, and alerted to the necessity of establishing a bargaining position, Asbury exhorted the Standard research organisation to diligence and speed in filing patents on all their discoveries. This effort undoubtedly paid off, even though some hastily written applications may have missed a point here or there.

For, despite the priority of Natta over American scientists in disclosure of polypropylene *per se*, there was room for significant discoveries that would improve the catalyst, the process and the product in many vital respects. In fact, it was only by the cumulative effect of a number of key discoveries of this sort that the original scientific discovery was developed into commercially successful processes and widely used products.

Given the large and numerous research teams that were soon deployed around the world, it was probably inevitable that most of these discoveries would be made sooner or later; if not in one group, then in another. Certainly, a number of the most important findings were duplicated more than once in various laboratories in both the U.S. and Europe. But in the race to the Patent Office, there are no prizes for 'place' or 'show', so timing, both in invention and in filing patents, proved to be of critical importance.

Standard's chemists made their share, or more, of important supplemental discoveries, winning some and losing some in terms of patent coverage. Like Shell and perhaps several others, they discovered that hydrogen was a useful modifier in the polymerisation reaction, but in the showdown at the patent office Hercules' Ed Vandenberg won out.

The most significant, and rewarding, contributions from Standard's research were novel methods of preparing and pre-treating catalysts to improve greatly their activity and specificity. The catalyst systems used by Ziegler, Natta and practically all commercial producers of polyolefins contain titanium in the trivalent state, usually the trichloride, generated by reduction of the tetrachloride. But the methods used in the reduction and subsequent treatments (heating, ageing, grinding) affect the crystal structure, particle size and activity of the finished catalyst. Interestingly, these different compositions have different colours, so the state of catalyst development in each laboratory at a particular time could have been judged by observing whether the latest catalyst to be investigated was black, brown, violet or purple*.

Tracing the origins, originators and respective merits of the innumerable variants is an intricate task beyond our scope, and will be left to other, more expert chroniclers (*see* the references in Ch. 1) and to interpreters of patent claims. We shall cite only as an example the contributions of the Standard research groups, since the catalyst variants they developed have been the most important and successful, as judged by patent coverage, widespread use and licensing by other users.

Major emphasis at Standard had continued to be on polyethylene through most of 1955, but propylene soon came to the forefront. Some of the earliest experiments were carried out in the group assigned to do product application and process development work. Since there was scanty product and only an embryonic process, they felt free to try out some ideas about catalysts. ('In the early days', one investigator recalls, 'everyone had an equal chance'.) C. W. Seelbach and J. W. G. McCulloch prepared violet titanium chloride by 'pre-reducing' the tetrachloride and, with such a catalyst, made solid blocks of polypropylene that proved to be hard and tough (even without solvent fractionation).

The swing to polypropylene was given momentum by William Sparks, the butyl rubber co-inventor, who by then had become director of research. He insisted that 'polypropylene is more important than polyethylene' and enforced his judgment by threatening to remove all ethylene cylinders from the laboratory. One of the last to be convinced was Art Langer, who felt that his exploratory program could make faster progress studying ethylene, since it was a somewhat simpler system. Eventually, however, even he was persuaded to work on polypropylene, and so the catalyst developments were attuned to the needs of that product.

---

*Behind the superficial differences in colour are significant differences in the extent of reduction of the titanium tetrachloride and the crystal form of the formed titanium trichloride, namely:

α (alpha)	close-packed crystals;
β (beta)	linear polymer;
γ (gamma)	cubic close-packed;
δ (delta)	randomly stacked crystal layers produced by cold working of alpha or beta forms.

Langer and Erik Tornqvist soon worked out the best proportions of ingredients and the best temperature to use in the 'pre-reduction' step. More importantly, they came to realise that the catalyst must be heterogeneous (two-phase) and that a finely divided catalyst should have the most active surface and should therefore give the best results. They then conceived two different methods of achieving the desired fine dispersion.

> Since crystals tend naturally to grow in size, the logical approach was to interrupt that growth, or to break the crystals down mechanically, or both. Langer and Tornqvist reasoned that if aluminium metal, instead of an aluminium alkyl, were first used to reduce titanium tetrachloride, aluminium chloride would be formed and would be intimately mixed (or co-crystallised) with the reduction product, titanium trichloride. Then the subsequent addition of aluminium triethyl would dissolve the aluminium chloride and cause dispersion of the titanium trichloride.

Tornqvist went to Seelbach and arranged for the use of his equipment to test this first idea: reduction with aluminium metal. The results were gratifying. A catalyst of unusually fine particle size was obtained and proved to be more stable and much more active than previous catalysts. Additional improvement was obtained by pursuing the second idea: grinding the catalyst (in a ball mill) to further reduce its particle size. In fact, the improvement was too great to be accounted for simply by reduced particle size, and it was eventually determined that a change in crystal form also occurred on intensive grinding.

The effects of grinding, as well as synthesis conditions, were also studied by Natta who, characteristically, carefully determined the crystal habits of all the various titanium trichloride preparations and christened them, 'alpha', 'beta', 'gamma' and 'delta' (see footnote, p. 121).

The specific, optimum preparation procedures developed at Standard were promptly patented but, unlike Natta's work, were not published until much later (1967, 1972!). Standard, in common with many other companies, had a publication policy that was very progressive as stated but very conservative as administered, since control was in the hands of the naturally cautious legal department. Their repeatedly imposed delays in publication were frustrating to the inventors, who thereby missed timely peer recognition. Such delays also probably retarded recognition and acceptance of Standard's catalysts.

Despite the company's reluctance to permit the research staff to publish their findings at scientific meetings or in professional journals, and despite Standard's relatively non-aggressive licensing policy, the merits of the finely dispersed catalysts eventually became recognised. And, once again, it was at an American Chemical Society meeting that a contact was made that had significant consequences.

In the autumn of 1957, Erik Tornqvist had been allowed to attend an ACS meeting, although not to give a paper there. Outside the meeting rooms, he was approached by C. C. Baldwin, a market development representative from Stauffer Chemical Company. Stauffer had been offering titanium trichloride for catalytic uses, but without great success. Stauffer were aware that their product was not as satisfactory as other proprietary catalysts, and were watching the patent literature for clues to improvements. Following his discussion with Tornqvist, Baldwin sent Standard a sample of Stauffer's trichloride and learned in return that Standard did indeed have superior catalysts of their own.

At a later meeting, Baldwin had a discussion with Art Langer. He cited Standard's just-issued Australian patents (which, as usual, were published well ahead of their U.S. equivalents), which disclosed preparation of titanium trichloride with aluminium as reducing agent. Baldwin offered new samples for Standard's evaluation that had similar characteristics. Eventually, these exchanges led to a proposal by Stauffer to license Standard's catalyst manufacturing process. But Standard chose not to license catalyst manufacturers, since the absolute volume of catalyst produced would never be large. Instead, they proposed to license companies that used the catalysts to make polymers.

Stauffer and Standard pursued their separate courses, and both were ultimately quite successful, Stauffer in marketing catalysts and Standard in licensing users and as a major producer of polypropylene.

But Standard did not achieve its position without trauma and travail. Lawsuits were required to establish the validity of Standard's patents and induce users to take licences. Internal difficulties also arose, due in part to the multiplicity of corporate entities involved. What we have been calling 'Standard' included the Esso Standard refinery at Baton Rouge, Louisiana, where a pilot plant was constructed for proving up a polypropylene manufacturing process based on the catalysts developed in the research program. All the research, however, had been done at the Linden, N.J., laboratories of Esso Research and Engineering Company. But it was Humble, then an operating affiliate of Standard Oil Co. (N.J.) that took the corporate bit in their teeth and plunged into large-scale production. ('They wanted a big venture', it was said.)

Coordination between these and other operating entities and the necessary staff functions was complicated and sometimes delicate. One observer likened the corporate meetings to the Vietnam peace talks in Paris, for the conference tables had to be very carefully set up in a way that would reflect the precise status of each of the various primary and subsidiary pieces of the corporate hierarchy.

With impressive, if questionable, determination to save time by short-cutting the steps normal to a major project, it was decided not to wait for definitive results from the pilot plant program. Accordingly, a larger, 'demonstration' plant was built quickly. It 'demonstrated' chiefly that often haste makes waste, for several items of inadequately tested equipment failed to work properly, and the plant never made good polymer. The full-scale commercial plant was designed and completed with little or no aid from the 'demonstration' plant. After the usual, inevitable start-up and shake-down difficulties, it operated satisfactorily and was later greatly expanded.

Polypropylene: a discovery whose time had come

Entirely apart from questions of scientific credits and the thorny complexities of patent priorities, it is obvious that the discovery of polypropylene was inevitable once Ziegler's polyethylene technology had become widely known. We have seen that at least half-a-dozen separate industrial or institutional laboratories made polypropylene, inadvertently or advertently, within the space of only a year or so, starting with nothing more than the knowledge of how to use Ziegler catalysts to make polyethylene. In several cases it took only a matter of weeks (or days!) to go from polyethylene to polypropylene.

Two qualifying comments need, however, to be added. (1) These were uncommon times. The mid-fifties constituted a period unique in the diversity and intensity of research effort directed to one particular field, and unique also in the resultant burst of fruitful creativity. (2) The unique and impressive aspect of Giulio Natta's contribution was his elegant elucidation and proof of the structure and morphology of the new high polymer he had discovered.

The prolific period at Milan Polytechnic

Although the companies who pioneered the commercial development of polypropylene had got their head start from advance information obtained through private contacts, it was not long before the full details of Natta's work became public knowledge through a series of publications that presented them with admirable explicitness and precision. Immediately after the polypropylene discovery, Natta completely reorganised the Institute program, putting all but a few of the staff of over one hundred to work on various aspects of stereospecific polymerisation. He deployed his forces across that field in a well planned, two-pronged attack that gave due emphasis to both scientific problems and commercial opportunities. Prof. I. Pasquon, who took his degree in Chemical Engineering under Natta in 1953 and now occupies Natta's former office as Director of the Institute of Industrial Chemistry, described to me the launching of that program as follows:

'Natta had always the intuition for the most important things—important from both the scientific and industrial points of view. After polypropylene was discovered, Natta immediately set forth three goals: (1) new plastics, (2) new fibres, and (3) new rubbers. These goals were recognised at the beginning and stated at the beginning in a very clear exposition. I remember he said what were the most important things we had to do after the polymerisation of propylene: we had to look at styrene and a lot of other monomers, and he said we have to prepare a polymer of butadiene, because that is the most important (*see* Ch. 10). He said all these things because he knew right away the problems of industrial chemistry and he knew very well the methods to be used.'

Natta must also have struck an optimum balance in research administration between over- and under-administration, for he drove his staff to an extraordinary level of productivity without their feeling driven or stifled in initiative. In fact, his associates uniformly characterise him as an easy man to work for—one who showed real interest in the progress of the work but left considerable freedom to the man doing it. In areas where his personal interest and familiarity were not the highest, he was content to delegate responsibility for daily supervision to others. However, there was not a formal hierarchy. Thus, Piero Pino was classified simply as assistant to Natta, although he was recognised as the 'number two man' on the staff on the basis of seniority and close association with Natta.

Working hours were not rigidly enforced, nor was the work tightly compartmentalised; senior people set their own schedules and were free to discuss their work with one another. But Natta visited the senior people every day in the laboratories and talked over techniques, results and ideas. This activity would usually occupy the entire morning, from 8 am to 1 pm, and would resume in the afternoon. Sometimes the discussions would be adjourned to the Director's home and continue there, even to midnight. Prof. Pasquon recalls staying at Natta's home for several days during 'holidays' when the discussions ran until 1 or 2 am.

Natta was constantly proposing new ideas, but he could still hear those of others. He never vetoed an idea outright, even when unconvinced, nor did he issue direct orders.

'I think this is important', he would say. 'If you can, please try it.'

The chemist knew, however, that there would be a follow-up to see how the test came out and to discuss its significance.

It seems obvious from all this that the chief factor in Natta's success as a research director was the intensity and determination of the man—a force of personality that was felt strongly by his people and swept them along the path he saw so clearly and pursued so unrelentingly.

The flood tide in research and publication

The results of this well conceived and well executed program stand as a landmark achievement in fruitful innovation. The formal scientific record of the first five years comprises no less that 170 papers (over 2000 pages), published by Natta and some three-score co-workers between December 1954, and December 1959. That pace—one paper on the average of every $1\frac{1}{2}$ weeks for five years—must itself qualify for the book of records. There is naturally some overlap among papers written for different audiences and published in different journals, countries and languages. But there is no serious 'puffery', and the scientific calibre of the papers is generally high*.

Robert Snyder, an American spectroscopist who spent a 'sabbatical' year at the Milan Institute in 1963, marvelled at the speed of oral and written communication. Commenting on the phenomenal output of papers, he remarked: 'Well, if they could write as fast as they talked, they had no problem. A man I knew would go home one night and come back the next day with a paper all written.' He also envied the speed with which approval for publication could be secured, in contrast to the ponderous mechanism of his home (industrial) laboratory in the U.S. In earlier years, however, each paper issuing from Natta's

---

*English translations or abstracts of the 170 papers have been collected in book form: *Stereoregular Polymers and Stereospecific Polymerization: The Contributions of Giulio Natta and his School to Polymer Chemistry*, Vols I and II, Eds Natta and Danusso, Pergamon Press, New York and Oxford.

laboratory was discussed in painstaking detail with the Director. Pasquon recalls once spending two hours with him to settle the precise wording of a single page.

Viewing this great output in perspective, the most impressive contribution of Natta and his 'school' is not the items mentioned in his own statement quoted at the head of Ch. 7 above, nor even the discovery of polypropylene itself. After all, the application of Ziegler's polyethylene catalyst to propylene was such a natural thing to try that someone was bound to do it, whether or not Natta did. In fact, as we have seen, it was done essentially contemporaneously in several laboratories. Nor did Natta originate the concept of stereoregular structures in polymers (*see* Ch. 2). The unique and outstanding feature of Natta's work is his elegant and precise elucidation of the crystal structure and chain configuration of stereoregular polymer molecules.

Thanks to their familiarity and capability in x-ray crystallography, as well as infrared spectroscopy, Natta and his staff were able to demonstrate immediately that their solid propylene polymer was microcrystalline, to measure the precise dimensions of the unit cell in the crystals and, from that, to make a rigorous deduction of the stereoregular nature and helical configuration of the polymer chain. The same approach was taken successively and successfully with each of the new stereoregular polymers they prepared from styrene, 1-butene, 4-methyl-1-pentene, etc.

What's in a new name?

The accepted terminology of conventional stereochemistry ('optically active', 'all-*D*', '*DDD*', '*DLDLDLDL*', etc.) proved awkward and inadequate to describe the many possible configurational variants of stereoregular high polymers. Fortuately, Natta had chosen well, even better than he knew, when he married a girl who was a language scholar and a semanticist. Rosita Beati Natta came to her husband's aid by coining an entire family of new names, based on Greek roots, which Natta used in his publications and successfully proposed for formal adoption*.

---

*At a dinner honouring Prof. Natta, Paul Johnstone of Hercules was seated next to Mrs Natta and learned from her the origin of the new nomenclature. He then had her translate for her husband's benefit his joking suggestion that the Nobel Prize should have gone to Rosita Natta, the 'real creator' of isotactic polymers. Natta did not seem to appreciate the joke.

A polymer with all substituent groups on one side of the chain was christened 'isotactic' (from the Greek words for 'same' and 'arrange in order'); one with regularly alternating substituents, 'syndiotactic' ('*syn-dio*' = 'together', 'two'); one with random order, 'atactic' (non-ordered). Other terms were generated as required for further structural variations as they were synthesised or postulated. The new terminology was a welcome help to scientific communication and perhaps an even bigger help in winning recognition of Natta's work as original and important. The hand that rocked the semantic cradle helped rock the scientific world.

The ubiquitous helix: spiral chains in natural and synthetic polymers

We digress briefly here to note a striking parallelism between the work of the molecular biologists, who so brilliantly elucidated the structure of important natural polymers (proteins, DNA) and that of the polymer chemists who are the principals of our story. Working independently and in different fields, the two groups evolved, almost simultaneously, concepts of structural regularity and spiral chain configurations in polymer crystals that are remarkably similar in principle.

As the threads of these two stories unravel, they occasionally cross or even entwine, and it is natural to ask how much one group may have benefited, directly or indirectly, from the other. As all islands are connected under the sea, so all high polymer research goes back to common beginnings—and in the beginning, of course, there were only natural polymers. But we can trace a much more direct connection, starting with Herman Mark, whom we have encountered so often in this narrative. Mark worked early and extensively in both fields—natural and synthetic polymers—and as early as 1928 (!) he proposed that in crystals of silk protein the polymer chains have a spiral shape. Another pioneering proposal was made by C. S. Hanes, who in 1932 suggested a helical model for amylose (starch).

Then Linus Pauling, holder of the Nobel Prize in chemistry and a dozen other prestigious awards, and an innovative (and controversial) contributor for many years to many fields, turned his attention from the fundamentals of chemical bonds to the unfamiliar and confused field of protein chemistry*. He began by unmercifully demolishing the 'cyclol'

---

*This marked shift of interest occurred after, and presumably because of, a serious (previously fatal) kidney disease that gave Pauling a 'vital' concern with anti-sera. It's an ill illness that does no-one good.

theory of protein structure (a sort of 'daisy chain' or 'chickenwire' structural scheme advanced by Dorothy Wrinch), and ended by proving the spiral chain structure which he named the 'Alpha Helix'. And when he published this structure, he made reference to Herman Mark's pioneering contribution on silk protein.

At least one other polymer pioneer, like Mark, was interested in both natural and synthetic polymers: Maurice Huggins. In 1937 he proposed spiral chain structures for a number of polymers, including polyisobutylene. That synthetic polyolefin had been discovered about five years earlier in the laboratories of I. G. Farben in Germany. Its discoverers (Otto and Müller-Cunradi) also concluded (1938) that it probably had a spiral chain structure.

Calvin Schildknecht, in his book on vinyl polymers, published in 1952*, described polyisobutylene and its supposed spiral structure, but he stopped just short of proposing the same structure for his own new polymer, crystalline polyvinyl isobutyl ether.

In the same year (1952), Watson and Crick were working with fitful feverishness to discover the secret of the structure of DNA, the all-important nucleic acid that is the key to the genetic code. According to Watson's own remarkably frank account†, their express goal was to discover and publish the structure of DNA before Pauling got it, and thus to snatch the Nobel Prize that otherwise might become Pauling's second‡. They took information anywhere they could get it, including the results of other workers at Cambridge∮ and data obtained covertly from Pauling's son regarding his father's unpublished work. Announced in 1953, their prize-winning discovery—the double helix structure for DNA—thus had roots that run back through Pauling to Mark.

The 'Year of the Double Helix' was followed immediately by the 'Year of the Threefold Helix'. For it was early in 1954 that Natta, Chini and Corrodini first made polypropylene and determined that its chains were coiled in what they termed a threefold helix (meaning, not that three chains were involved, but that it takes three monomer units

---

*Schildknecht, *Vinyl and Related Polymers,* Wiley, New York (1952).
†Watson, *The Double Helix,* Atheneum, New York (1968).
‡Pauling did eventually win a second Nobel Prize—the Nobel Peace Prize!
∮Compare Watson's account with that in Sayre, A., *Rosalind Franklin and DNA,* Norton, New York, (1975).

to complete one turn of the spiral chain).

The near-simultaneity of the two discoveries was apparently mostly coincidence, for one gets the impression that the research races on natural and synthetic polymers were being run on parallel tracks with much less joint awareness and contact than one might expect. Certainly, there was far less mutual reinforcement than might have occurred had not those who were taking apart complex natural polymers and those who were putting together the long but simpler molecules of synthetic high polymers considered themselves in separate worlds*.

But to Nature, all things are natural, even when man acts as her agent in making them. So it should come as no surprise that the spiral shape that provides the most comfortable fit for like chain segments in proteins and in DNA turns out also to be the natural habit for polypropylene and other synthetic polyolefins.

Indeed, it probably came as no surprise to Natta, who had a predilection for noting similarities in nature ('*Natura non facit saltus*'—'Nature does not make jumps'—he was fond of saying). With remarkable swiftness he perceived, proved and published the helical structure of stereoregular polyolefins. A lasting, if somewhat flamboyant, testimonial to that insight hangs in Natta's office: an imaginative painting that shows the scientist ascending a spiral staircase consisting of an enormously enlarged helical molecule of polypropylene.

Natta in the New World

Thanks to his publications and publicists, Giulio Natta achieved almost overnight fame, and he was soon in high demand for visits and lectures. On these occasions he was lionised and beseiged by admirers, some genuine and some using admiration as a tool in seeking free information or an inside track to a licence.

In 1956, as a fellow Ligurian from Genoa had done centuries before, Natta landed in the New World and found the natives friendly. In fact, the attentions were so intense and competitive as almost to constitute

---

*As with a few notable exceptions, they still do. Item: the highly readable and insightful review by Olby, 'The Macromolecular Concept and the Origins of Molecular Biology', *J. Chem. Educ.*, 47 No. 3, 168 (1970) shows no awareness of the work of Mark, Huggins, Flory, Schildknecht, Natta or even Wallace Carothers.

body-snatching. Through the indefatigable Prof. Mark, he had been invited to lecture at the Gordon Conference on High Polymers, as had Calvin Schildknecht some nine years before (Ch. 2).

Natta arrived with an entourage that included his wife, his associate, Dr Piero Pino, his sponsor, Giustiniani, and Mario Ottolenghi, who was to head Montecatini's U.S. subsidiary, the Chemore Company. Montecatini was thus prepared both to pursue opportunities and defend their scientific golden goose from would-be poachers.

Gordon Conference attendance was normally limited to about a hundred at any one session, but on this occasion the conference management was pressured into accepting over twice that many, including many corporate vice presidents and directors of research. All were anxious to meet and hear Natta and, particularly, not to be out-manoeuvred by their competitors. They laid immediate and heavy siege to the distinguished visitor, but they quickly found the language barrier formidable, even with Mrs Natta and some of the other Italians serving as interpreters.

When the time came for Natta's lecture, he delivered it to a packed room, speaking in 'English', but his accent was so strong and his slides so hard to decipher that his listeners were unable to get much from it. After the lecture, most of his entourage left the conference, and most petitioners had become discouraged, so Natta had more time to himself.

There was one American who was able to take advantage of this situation, partly by virtue of speaking Italian fluently. Aldo DeBenedictis, a research chemist at Shell Development Company who had followed Natta's publications and was himself doing research on stereospecific polymerisations, was one of those who had come to the conference hoping to meet Natta in person. Thwarted by the throngs, he resorted to writing a note to Natta in Italian, and was rewarded with an invitation to meet and talk with both Nattas in their room.

As the conference went on, DeBenedictis saw more of Natta, who was happy to have someone to talk to in his own tongue. One evening, DeBenedictis was asked to serve as interpreter for Natta at an impromptu question-and-answer session. He was happy to do so, but was somewhat non-plussed when Natta began answering questions in French instead of Italian; nevertheless, he was able to handle the tri-lingual exchange to everyone's satisfaction. In this way, a friendship developed that was maintained through subsequent visits and correspondence.

Prof. Natta, still the chemical engineer, expressed strong interest in Shell's unique synthetic glycerine process, and DeBenedictis, at the sacrifice of his own vacation, agreed to take him to Houston to see the plant. Natta then agreed, on returning to New York, to spend an evening with top Shell Executives at the Fifth Avenue penthouse apartment of the president of Shell Development Company.

In the amiable ambience thus created, Natta hinted to DeBenedictis that it might be possible to arrange an exclusive U.S. licence for certain types of polymers (probably he had in mind the rubber-like ethylene-propylene co-polymers). This was passed on to Shell management but, like others faced with similar early opportunities, Shell passed up this opening and so had no tangible result from the potentially fruitful *entente cordiale* that DeBenedictis had created for them. Much later, after developing their own polypropylene process, Shell joined the procession of those who went, cap in hand, to Milan, and settled for purchase of non-exclusive rights under whatever polypropylene patents Montecatini might eventually get.

Other visitations and other high-level discussions filled Natta's busy American schedule. From the Gordon Conference, he had been flown off to Bermuda in a private plane by an executive of one of the 'Big Four' rubber companies for discussions with (and attempted persuasion by?) their top management. However, if the possibility of an exclusive licence was brought up on that occasion, it must have been passed over also.

On the Nattas' second visit to the United States a few years later, they were received in an equally cordial but more orderly manner, and he was somewhat less of an object of frantic pursuit by industrial opportunists. A seminar at Brooklyn Polytechnic Institute became an occasion of historic as well as scientific significance. Prof. Mark had operated for many years a series of 'Saturday Seminars', which wags dubbed the 'Saturday Subway Science Series' because the invitations always gave directions for reaching BPI via the New York subway system. On a Saturday afternoon nearly a decade earlier, Calvin Schildknecht had lectured at a seminar on 'Isomerism in Macromolecules', chaired by his former professor and mentor, Maurice Huggins. Now Mark asked Schildknecht to preside at the seminar at which Natta was to lecture on stereoregular polymers.

Rosita Natta was concerned with whether or not her husband's slides would be legible and, although she did not say so, probably also with whether or not he would be understood. Her concerns were justified, but the occasion was still a success. Mark opened the proceedings by introducing Schildknecht as the first to suggest a stereoregular structure for a specific polymer. He then brought Giulio Natta to the platform and introduced him to Schildknecht. In his introductory remarks, Natta politely acknowledged Schildknecht's contribution.

Other stops on this tour included lectures at Notre Dame and at San Francisco, where the Nattas were met by Aldo DeBenedictis. Natta confided his concern about the quality of the English translation of the paper he was to read and invited comments. DeBenedictis, aware that the awkwardness of the transliteration would be compounded by Natta's pronunciation, suggested that the paper be read by someone else, with Natta on stage. But the Italian's pride would not let him adopt this course, and he made the presentation as best he could without assistance.

Meanwhile, back at the patent office

Natta's prolific and highly visible activities on the scientific front did not mean that he or his sponsors were neglecting the practical and commercial aspects of his discoveries. In contrast to Ziegler, however, Natta stood to receive no direct share in any royalty income and was happy to leave the drafting and filing of patents of Montecatini's lawyers. The company was certainly diligent in ensuring full patent coverage and pilot plant development work. DaVarda, the lawyer who headed Montecatini's in-house patent and licensing work in this field, made it his career from then to retirement. In addition, well known legal firms were recruited in the U.S. and other major countries to prosecute Montecatini's cases outside Italy.

The flood of patents stemming from Natta's work rivalled the output of scientific papers. Just the most significant of the patents assigned to Montecatini total several dozen. The iceberg of which this is the visible tip was enormous, and the total cost to Montecatini in filing and prosecuting patents, interferences and licensing activities over two decades must amount to tens of millions of dollars. Doubtless, it was money well spent; Montecatini's net income from royalties must be a much greater sum, and may even exceed the profits on their own manufacture and sale of stereoregular polymers.

That effort was multiplied manyfold around the world as each company developed and patented its own variations on the basic catalysts and processes of Ziegler and Natta. Montecatini's licensees did not receive very extensive process information, and most companies chose to do their own process and product development while waiting to see whether Montecatini would end up with dominating patents in their country, and how aggressively they would enforce them. The total of all patents in the field, worldwide, numbers in the thousands; creating and defending them has made careers for scores of lawyers and clerks.

Many chemical engineering problems had to be solved in laboratories and pilot plants before commercial plants could be designed and built. Chief among these were extraordinarily stringent purity requirements and unusual safety problems. Handling the volatile and flammable gases, ethylene and propylene, was old hat, but slurries of catalysts that exploded in contact with water and burst into flame if touched by air were something else.

Paolo Chini, one of the original three Montecatini representatives sent to Ziegler's laboratory, had an accident there with aluminium triethyl, and the resultant fire sent him to the hospital for weeks and left him with permanent scars. Later, Montecatini experienced at least one serious fire in their own pilot plant that destroyed important records as well as equipment.

At least one commercial plant was built (by another company) with full expectation of having a fire every time the catalyst was charged, and was designed not to prevent but only to contain such fires. At one time it was said that every commercial plant using Ziegler/Natta catalysts had had at least one fire fatality. Gradually, however, the process problems were brought under control and excellent safety and production records were achieved.

Giustiniani's Roman triumph in Milan

It was a time of triumph for Montecatini and its head man. The Italian scientist sponsored by Giustiniani had won for Italian science and for the Italian chemical industry their coveted place in the sun. Executives of the world's top chemical companies (with certain conspicuous exceptions) undertook pilgrimages to Milan, accompanied by licensing experts, chemists and/or wives, according to whichever they thought would best serve them on the Italian scene. They were politely received and allowed to present their credentials and petitions in accordance with a scenario scripted by Montecatini, probably by Giustiniani himself.

If the visitors were of less than top rank, they got to see only DaVarda and Orsoni, who would explain that Giustiniani had delegated full responsibility to them for licensing matters. A morning negotiating session might get some specific proposals on the table, but not a decision. The visitors would then be treated to the fabled long Italian lunch and afterwards would be graciously invited to spend an hour resting before resuming discussions. The Italians would reappear at the conference table with their minds suddenly made up about the morning's proposals, probably by virtue of having consulted the boss during the 'rest' period.

But if his rank and title were sufficiently impressive, the visitor might be ushered into the presence of the Great Man himself. The entrance at the end of a huge office and the long approach to the area occupied grandly by Giustiniani's desk were apparently calculated to foster the image of a petitioner humbly approaching the throne. If things seemed to go well, the meeting might end with fulsome assurances of friendly accord and of Giustiniani's ability to settle everything to everyone's satisfaction. This created a climate in which it would have seemed impolite to insist on a signed piece of paper. Later, however, it might turn out that circumstances had somehow changed, and Giustiniani's mind with them.

He had a special relationship with the licensing manager for the Shell companies outside the U.S., Adrian Koeleman. They had become acquainted before World War II during lengthy negotiations on the oil side of the business. Now, in the mid-fifties, Shell Chemicals Ltd had acquired Petrochemicals Ltd and with it the exclusive Ziegler licence for the U.K. Royal Dutch Shell were interested in manufacturing polypropylene (and, later, polybutadiene) on the Continent as well as in England. They had the dominant patent position in England through PCL but, due to Ziegler's oversight, it did not extend to polypropylene. Montecatini had the dominant patent position on the Continent, but lacked full resources, particularly marketing capability, to exploit their position. Each company needed something the other had.

The result was a long-drawn-out period of negotiation, during which Giustiniani thought nothing of cabling Koeleman in London on Friday asking him to come to Milan for a meeting on Saturday. It got to the point where Koeleman seriously considered getting himself an apartment in Milan so he could respond to these repeated summonses with less inconvenience.

Giustiniani was a big man physically, and he tried to live up to the image. 'All Italians are born Signori', he would say. A problem related to this attitude was his well-reported reluctance to delegate. This is not unusual in Italy, where it is said, 'Every company is operated as a proprietorship', but Giustiniani carried it to an extreme which was unworkable in a company the size of Montecatini.

It was his boast that he had no less than thirty-eight people reporting directly to him, and he added, with satisfaction, that they all had to stay in their offices as long as he was in his. This was essentially true, although his underlings did organise, in self-defence, an 'early warning' system that enabled those unlikely to be called on to slip away home as early as 9 pm instead of 10.30 or 11 pm. They also escaped for three weeks in August, when the whole company closed down for vacation, but even then Giustiniani could be found in his office. He concerned himself with everything and tried to settle everything himself. ('A meddlesome busybody', he was called by some.)

These characteristics meant that dealing with Montecatini was difficult and tedious. It was rendered doubly so by the team of DaVarda and Orsoni. DaVarda prided himself on meticulous detail and was the perennial objector—the one who at the last moment would always insist on something more for Montecatini. This became so irritating that there were times when the negotiators would be on their feet, pounding the table and threatening each other physically, until cooler heads prevailed. As we have noted before, this may have been partly an act, DaVarda playing the 'bad guy' and Orsoni the reasonable peacemaker. But whatever position and strategy he might adopt on a given occasion, they were always designed to further his high ambitions for himself, Montecatini, and the Italian chemical industry.

In this environment, the negotiation of a multinational, multiproduct joint venture and licensing deal took Shell two and a half years to complete. Other negotiators were able to consummate less involved deals in less time, but none has been heard to say it was easy. Yet those who were determined persevered, for there was no other well from which to draw. In contrast to the situation in linear polyethylene, for which at least two other processes were available for licence (from Phillips and Standard of Indiana), only Montecatini offered a polypropylene licence.

Both Phillips and Indiana had worked early and hard on propylene (in fact, Hogan's first experiment, as we have seen, was with propylene), and those who believed patent claims and rumours were expecting commercial processes to emerge from one or both of these sources at any time. But in this, as in so many instances, the step from patent to production proved so big that it has still not been made. Giulio Natta, who was so sure that 'Nature does not make jumps' and therefore that a catalyst that polymerises ethylene is bound to polymerise propylene also, might have had some difficulty accounting for the apparent high specificity of the Phillips and Indiana catalysts.

Nor were things much easier on the technical front. Those who paid to get access to Montecatini's technical and process information were sometimes disappointed in the calibre of the process engineering reflected in the engineering design and operating standards. Although the Italian firm would have heatedly denied it (after all, Natta and Giustiniani were both chemical engineers!), this deficiency doubtless contributed to the extensive replication of process research and engineering in other laboratories and engineering offices, and thus to Montecatini's difficulties in continuing to play King of the Hill.

It also resulted in many improvements in both product and process, as the rapidly mounting body of patents testifies. Some of these 'supplemental' discoveries proved of absolutely vital importance in converting the laboratory reaction into a practical commercial process. Two examples have already been described: the discovery by Standard (and others) of the merits of 'purple' titanium trichloride in catalyst preparation (still further improved by fine grinding), and the discovery, first made by Hercules' Ed Vandenberg and later by others, of the salutary effect of hydrogen in controlling the polymerisation reaction (*vide supra*).

One other example will be cited from personal experience. It was common knowledge that water destroys Ziegler/Natta catalysts and that reactants must therefore be scrupulously dried before use. But it was discovered in the Shell Development Company laboratories that drying could be overdone. An absolutely anhydrous system did not give as fast or consistent polymerisation as one with a judiciously tiny trace of water present.

This is the sort of finding that is unlikely to be made while working on laboratory bench scale, because routine procedures will not remove the last few molecules of water. But with the larger volumes of development-scale work, drying is more apt to be complete and even a measurable amount of water becomes only a trace, so the effects of overdrying are more easily observed. Once it was recognised that water is a useful control agent, techniques were worked out for injecting the critical traces needed*.

*See page 138.

Other useful additives (e.g. amines) have also been found that give similar benefits.

These and many other improvements had to be made before a smoothly operating *process* could be achieved, but equally numerous and important innovations were required to ensure that the process produced a practically usable *product*.

Finding jobs for polypropylene

The classical question about the newborn baby is appropriately asked about any new polymer: 'What is it good for?' In both cases, laboratory tests can promise a fair future but cannot predict a specific career. Successful introduction of a new polymer always requires diligent work by application research and market development groups. In the case of polypropylene, it appeared for a time that the traumatic experiences with linear polyethylene were destined to be repeated. The early testing suggested that while the concept of a hard, high-melting polymer from propylene was exciting, as a practical plastic it might not prove much good for anything. As with Ziegler and Phillips polyethylenes, high crystallinity made solid pieces translucent or opaque and caused warping in mouldings. Polypropylene also displayed its own peculiarities in moulding behaviour (due to sharp melting point and compressibility in the molten state) that caused some experienced moulders to condemn it at first for injection moulding or extrusion applications. Unlike polyethylene, it became brittle at only moderately reduced temperatures. Most alarmingly, it displayed a dramatic degree of instability when exposed to heat, light and air.

These multiple crises made for some frantic times in the laboratories. But the research teams did not panic utterly. The wiser heads knew that all new plastics, like newborn puppies, are highly vulnerable until they have received their inoculations. With rare exceptions, high polymers are highly unstable: they are inherently prone to oxidise, ozonise, hydrolyse, re-crystallise, de-polymerise, become insoluble, discolour,

---

*I happened to be in Europe when I got word of this discovery. I promptly mailed to Harry Cheney, the inventor, a miniature replica of Brussels' famous statue of the little boy 'making water', with a note saying that it might suggest a means of introducing the 'magic ingredient'. Cheney had the figurine mounted at an appropriate spot in the pilot plant, but other, more practical means were found for doing the actual injecting.

embrittle, or suffer various combinations of these and other, subtler afflictions. Hence, each new polymer must be studied to reveal its major weaknesses and to find means to mitigate them. This may involve modification of the polymer itself or blending it with appropriate stabilisers and other additives, or both. Polypropylene was neither an exception nor an extreme case; the deficiencies noted above simply meant that special modifications and formulations had to be worked out for each major use. All these product and process problems were attacked on many fronts in many laboratories, and by one means or another were gradually overcome.

Concurrently, by what might be termed a rational trial-and-error process in the laboratory and in the market place, polypropylene was learning just what applications it could fill better than anything else. As with all new materials, polypropylene could not instantly create entirely new uses for itself; its applications and markets had to be won away from older materials by proving it could do a given job better and/or more cheaply. Other plastics, metals, wood, paper, ceramics and natural and synthetic fibres were all fair game. These contests had some surprising successes and some embarrassing setbacks.

The first thought of those charged with marketing linear polyethylene had been that it would simply displace high-pressure polyethylene because it was harder, stronger and higher melting, but its greatest utility turned out to be in uses formerly served by metals or glass (e.g. containers). This lesson had to be learned anew with polypropylene; it was touted first as a competitor for linear polyethylene because it was still harder, stronger and higher melting. But again, the ultimate use pattern developed mostly in other areas.

An early bright hope for polypropylene was in thin films used for packaging. It could be extruded and stretched ('biaxially oriented') to produce sparkling clear films that rivalled cellophane. In fact, one of Orsoni's exuberant predictions to Giustiniani was that so much cellophane and paper would be driven out of the market that the companies in those businesses would be forced to the wall. Of course that did not happen; in fact, polypropylene's film uses have grown much less spectacularly than many others.

Fibre and fabric uses presented other surprises. On the borderline of acceptable heat resistance for use in fabrics (too low-melting for safe ironing) and, initially, of acceptable stability, polypropylene experienced a chequered adolescence before finding its proper assignments in the fibre field. One company (a subsidiary of Standard Oil Company of

New Jersey) spent two million dollars on promoting polypropylene fibre for women's stockings, on the strength of its greater snag resistance and lower cost compared to nylon (does anyone remember 'Vectra' hosiery?), only to find that women would not buy it a second time because the stockings were 'baggy'.

This was one fiasco that could have been avoided by a thoughtful look at laboratory test data. The clue was to observe the rather sluggish recovery of polypropylene fibres after bending, and for that purpose looking at individual fibres bent around a rod would have been less expensive and scarcely less revealing than looking at stockings bent around women's knees, even if less interesting.

The same limited resilience tended to limit the utility of polypropylene fibres in carpets, upholstery and apparel. But, to the surprise of some of us who had worried with the problem of inherent instability to sunlight, heat and oxygen, outdoor-indoor carpeting of adequate stability was eventually developed and this application proved to be one of the major product successes. Upholstery uses are also now well established. In England, Imperial Chemical Industries concentrated their formidable forces on the development of a variety of fibre applications, with considerable success.

Despite all problems and setbacks, the total of all uses continually grew at a rate that surprised those too close to the problem trees to see the forest of opportunities. In fact, polypropylene established an early pace that, if sustained (and, in fact, it has been), will set a new track record as the fastest growing polymer in all the history of plastics. Forecasters are already looking ahead to the day when this latecomer may take over the volume leadership from all other contenders.

These scattered samplings of skirmishes on the application and marketing fronts serve to remind us that the inventor is the indispensible initiator, but that for his invention to become a case of 'fruitful innovation' there must ensue a long chain of subsequent events and an equally indispensible succession of innovations by those responsible for each link in the chain.

10 Fruitful Innovation — 3. Synthetic Rubber

It is not strange that a material having the properties of natural rubber should arouse the curiosity of man and tease his imagination.
–Erik Tornqvist

The challenge

The marvellous high elasticity of rubber makes it one of man's most useful materials and at the same time has presented him with one of nature's most intriguing secrets. For over a century men tried but failed either to explain or to duplicate the unique properties and composition of tree-grown rubber. It is still true that 'only God can make a tree'; until recent times, only the rubber tree could make rubber.

It was long known that rubber was some sort of a polymer of isoprene, a five-carbon diolefin:

$$CH_2{=}CH{-}C(CH_3){=}CH_2 \qquad\qquad CH_2{=}CH{-}CH{=}CH_2$$

isoprene — butadiene

However, all efforts to convert it or the simpler, four-carbon diolefin, butadiene, into rubbery polymers gave only materials with poor elasticity and strength. The synthetic rubber that shod the war machines of both Axis and Allied forces in World War II was not even a pure diolefin polymer, but a random co-polymer of butadiene and styrene:

$$\cdots(CH_2{-}CH{=}CH{-}CH_2{-}CH_2{-}CH(C_6H_5){-}CH_2{-}CH{=}CH{-}CH_2)_n\cdots$$

'Buna S'; 'GR-S'; 'SBR'

This has an irregular and branched-chain structure. Developing the process and production capacity for this rubber in time to meet wartime need was a technological triumph, and even today the world uses more of this synthetic than of natural rubber; nevertheless, to chemists it was still an inelegant and crude approximation of the ideal.

Just how crude became apparent in the 1940s, when it became generally known that natural rubber is another example of nature's exquisite specificity: a linear, long-chain polymer of isoprene units linked exclusively in head-to-tail and *cis*-1,4- configuration, i.e. with the CH_2 groups of the polymer chain always on the same (*cis*) side of the double bond:

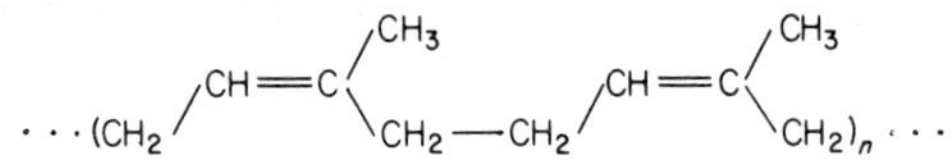

The isomeric structure, which differs only in that the CH_2 groups lie on opposite sides of the double bond, is not a rubber at all, but a tough resin (balata, gutta percha):

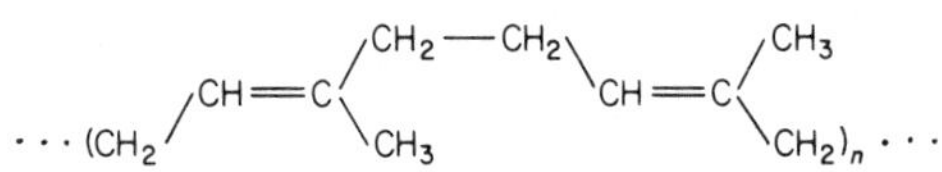

It was now obvious that no random and irregular synthetic polymer could hope even to come close to matching the precisely controlled structure of 'tree-tailored' rubber. For a successful synthesis, not only would thousands of isoprene units have to be added together in each molecule in perfectly linear sequence (no branching), but the units would have to be joined in only one of at least *eight* possible ways (or twelve, counting stereospecific structures generated by *D–L* isomerism*). The requisite degree of control over the polymerisation reaction was far above anything that had been achieved with any known catalyst or reaction conditions. Consequently, most chemists (certainly including this one) despaired of ever duplicating the natural rubber molecule.

---

*1,2-, 3,4-, 1,4-cis-, 1,4-*trans*-; each in either head-to-head or head-to-tail configuration. The 1,2- and 3,4- isomers can also be isotactic, syndiotactic, atactic, etc.

But, fortunately, not all.

It was in keeping with the historical position of European, and particularly German, scientists as the pioneers in synthetic rubber research that Karl Ziegler displayed early interest in polymerising diolefins and that Giulio Natta and the Montecatini Company viewed synthetic rubber as a major research and commercial goal. Despite this tradition and head start, however, by 1950 the prime locus of serious research on synthetic rubber had shifted to the United States. This was a direct result of the momentum generated by the enormous wartime effort that built the huge industry from scratch and in the process created a major research capability.

The Goodrich story–number two tries harder

As a legacy of the wartime pooling of know-how and talent that made the whole achievement possible, research information in the synthetic rubber field was still being exchanged in the 1950s among industrial and academic laboratories under government sponsorship and financial support. Several companies had tried various approaches, some within and some outside the 'pool' agreement, to duplicate natural polyisoprene synthetically. The B. F. Goodrich Co., one of the American 'Big Four' rubber manufacturers, even had a group working on a laborious, multistep chemical reaction sequence whose last step, they hoped, would yield a sample of *cis*-polyisoprene, however miniscule. Their hopes did not extend to any thought of a practical process based on such an approach, but it was government money being spent, and it would have given them a scientific 'first'. Firestone, also working within the 'pool', was re-examining the alkali metals as polymerisation catalysts, and had already obtained positive results, as we shall see later. But the total American research effort that had any potential for a successful duplication of natural rubber was neither very extensive nor very seriously regarded.

This situation in America was basically altered, even though that was not apparent at first, by an event in Europe. That event was Karl Ziegler's discovery of linear polyethylene.

The coal and chemical companies that supported Ziegler's institute were, of course, kept informed of significant developments, and thus knew about the polyethylene discovery before anyone outside Germany. But one of these companies, Ruhrchemie, also had an information exchange agreement with a U.S. company, Goodrich-Gulf Inc., a 50:50

joint venture by B. F. Goodrich and Gulf Oil Company in chemicals and polymers. Thanks to this agreement, a Goodrich-Gulf representative learned of Ziegler's discovery while visiting Europe in late May 1954, and the next day their representatives were on Ziegler's doorstep in Mülheim.

By July, an option agreement had been signed giving Goodrich-Gulf one month in which to examine samples of Ziegler polyethylene and reach a decision about a permanent licence. The price was modest (reportedly, $50000), but one of the conditions was that the recipients agreed not to analyse the polymer to find out what kind of catalyst was used (as they could easily do by burning the polymer to concentrate traces of catalyst residues and analysing the ash). Here, again, Ziegler accepted a personal commitment he had no way of policing, but this time his confidence was not abused. Frank Schoenfeld, the Goodrich-Gulf vice president who signed the agreement, made it clear to the research team that this restriction was to be honoured, and it was.

B. F. Goodrich had an excellent research facility and some good polymer chemists at Brecksville, near Akron, Ohio, and Goodrich-Gulf, spurred by the short fuse on their option, lost no time arranging to get a laboratory program started there. They were soon convinced that the properties of the new polyethylene were interesting enough to warrant exercising their option for a licence, and Ziegler was notified accordingly. Ziegler tried to persuade Goodrich that they should take an exclusive licence for the U.S., but they demurred, saying that the potential importance of the discovery was too great to confine it to just one company. Translated from the language of negotiation, that meant they were not convinced it was good enough to be worth the extra price for an exclusive.

In the long run, Goodrich's statement proved to be more valid than they anticipated, and Ziegler had repeated cause to be grateful for his failure to sell an exclusive licence to either Goodrich or Hercules Powder Co., who had even let their option lapse. Enough other companies took non-exclusive licences to swell his total royalties (and the total market for the products) far beyond what a single exclusive would have accomplished.

A very short time (about two months) after Goodrich got full information and rights to use his catalyst system, Ziegler was startled to receive a transatlantic telephone call from Dr Waldo Semon, Goodrich's director of polymer research, advising him in confidence that they had used that catalyst in a successful synthesis of *cis*-1,4-polyisoprene that essentially duplicated the structure and properties of natural rubber.

Ziegler probably received this news with mixed emotions. He was, and remained, grateful to Semon for giving him such prompt notification, but at the time his dominant reaction must have been chagrin. He was still tasting the bitter pill of losing polypropylene to Natta and now, once again, he had to accept the fact that someone else had used a Ziegler catalyst to make an epochal discovery that he, Ziegler, might well have made, given a bit more time and a bit less preoccupation with ethylene.

The other common diolefin, butadiene, was the first monomer Ziegler had polymerised—back in the 1930s—and he surely would have come around fairly soon to rechecking butadiene and going on to isoprene. In fact, Heinz Martin had made a new but somewhat casual test with butadiene in January of 1954 and had obtained a small amount of rubbery polymer. The conditions used, optimum for ethylene, were non-ideal for butadiene, and his yield was low. He did not examine the product for crystallisability or attempt to determine its *cis*-1,4- content. Similar experiments with isoprene had given poor results.

The failure to get any better results at Mülheim is probably due to two factors. One was uncertain purity of monomer, a chronic problem that was underestimated and never completely conquered. Goodrich were more fortunate in having isoprene of relatively high purity available. The other factor, at least in Ziegler's earlier work with isoprene and butadiene, was his practice of using ether as a reaction solvent. This was a natural choice in classical organic chemistry but, unfortunately, ether interacts with the Ziegler polymerisation catalyst to impair its stereodirecting ability. Ziegler had no way of knowing or anticipating this, and his results suffered accordingly.

The public announcement in the U.S. came on December 2, 1954. In a press release, the president of Goodrich-Gulf said: 'American scientists have finally succeeded in reproducing the molecule of crude, or tree-grown rubber, a major scientific achievement . . . a goal of world scientists for generations.' The discovery was credited to a research team 'assigned' by Goodrich-Gulf to the Goodrich Research Center in Brecksville, but the discoverers were not named, nor was there any mention of Karl Ziegler.

The Wall Street Journal, The New York Times and many other major newspapers and trade publications gave full coverage to Goodrich's announcement and offered their own editorial comments. The Akron (Ohio) Beacon-Journal spoke for many when it asked, after hailing the scientific achievement: 'Who were the individuals who collaborated in this important discovery? . . . here is one American team which deserves honour.' But it was to be almost a year before this question was answered in full and in public.

Reactions to the press announcement were highly varied. Competitors were spurred to re-examine their own positions and policies, leading in one case even to a nearly unprecedented pooling of forces by two major rivals, as we shall see later. Others in the rubber industry adopted a 'wait and see' attitude. Research people who were privy to the 'pool' research reports remembered reading of Goodrich's earlier, laboured, chemical synthesis approach and, confident that nothing of greater significance would go unreported or be withheld from the pool agreement, were puzzled as to what the fuss was all about.

The rest of the chemical industry was generally impressed by the scientific achievement but not by its economic significance. My own reaction was probably typical. The president of my company (Shell Development Co.) sent a clipping of the press announcement with a note to me (his technical assistant) asking: 'What does this mean to us?' I replied, with all the assurance of the expert advising the uninformed executive, that the duplication of natural rubber's structure was a milestone in science and might even win its discoverers the Nobel Prize, but that it would probably never lead to a commercial process and would remain just a 'laboratory curiosity'. Neither the Nobel Committee nor my own company took my comments much to heart, and only a few years later Shell Chemical Company became the first to manufacture synthetic polyisoprene on a commercial scale (although not with a Goodrich or Ziegler catalyst).

The Goodrich discovery was so impressive, so appropriate and so quick as to invite citation as a textbook example of successful application of the classical scientific method, with swift and sure progress from concept, to hypothesis, to experimental verification, and to triumphant conclusion. Nothing in the official announcement or first publication indicated otherwise. But, in fact, it was quite otherwise. The result obtained in the key experiment was not its original goal; a strong element of fortuity was present. This is hardly surprising; those with experience in both fields know that the course of true research is no smoother than that of true love, and that research, as J. C. Swallow pointed out (Ch.2), is always more orderly in the literature than in the laboratory. It is illuminating, therefore, to trace the actual course of events at Brecksville.

Horne's happy accident

Sam E. Horne, Jr, had come to the Goodrich Research Center only four years earlier (1950) with a fresh Ph.D. in organic chemistry from Emory University. His thesis work was on 'natural products', but his only exposure to polymers had been one prosaic lecture course in graduate school (one more than most of his contemporaries had!). At Goodrich, as at other industrial research laboratories of that time, necessity had mothered the custom of making polymer chemists out of organic chemists by on-the-job experience. A favourite starting point was the now-classical systematic study of Dr Frank Mayo and co-workers* on co-polymerisation; measuring the ratios in which different monomers choose to combine under standardised conditions of polymerisation in aqueous emulsion, using free-radical catalysts. So Sam Horne's initial assignment was to repeat and extend this type of fundamental but outwardly unexciting measurement.

The research work at Brecksville was not rigidly programmed, and the chemists had reasonable freedom in choosing how to attack their problems. ('It's still pretty informal', Horne says, 'only now there are several more levels'.) But the laboratory was no ivory tower. Goodrich is like most industrial organisations who dutifully intone the litanies of reverence to the Diana of 'pure' research but lay their heavy gifts on the altars of the *lares et penates* of 'goal-directed' and 'applied' research. Thus each of Horne's subsequent assignments had a specific practical goal and dealt with a plastic or rubber technology in which Goodrich already had a commercial interest. When the Goodrich-Gulf option precipitated a crash program on Ziegler polyethylene, Horne was one of those assigned to it.

Following Ziegler's procedures, he quickly confirmed Ziegler's results, but this was not enough; it was important also to learn how much the scope of this new chemistry might be extended. Being by now an 'old rubber chemist', Horne felt that the relatively good heat resistance of Ziegler polyethylene might be further improved by making it 'vulcanisable', i.e. capable of being crosslinked. By analogy with rubber (particularly butyl rubber, in which co-polymerisation with

*Mayo, F., *et al.*, *J. Am. Chem. Soc.*, **66**, 1594 (1944) and *J. Am. Chem. Soc.*, **70**, 1521 (1948).

small amounts of isoprene introduces enough unsaturation to permit vulcanisation with sulphur), it was logical to try co-polymerisation with diolefins. And so, even though Ziegler had told him that diolefins would not polymerise with his catalyst (!), Horne tried a mixture of ethylene and isoprene. He did get some polymer, which he sent to James Shipman in the spectroscopic laboratory for infrared analysis, hoping for some evidence of co-polymerisation.

Before long, Shipman called Horne on the phone to complain.

'Sam,' he said, 'you're trying to pull my leg. You've sent me a sample of natural rubber.'

After explanations and reassurances that the sample was legitimate, it was put through a quick separation procedure which showed that it was no co-polymer, but a mixture of polyethylene and *cis*-1,4-polyisoprene. The next step, obviously, was to try polymerising isoprene alone. This, too, succeeded.

'After that', Horne recalls, 'all Hell broke loose!'

It is worth noting here the several critical circumstances that were the keys to Horne's success:

(1) Isoprene of high purity was available to him because another group had been trying to make an isoprene analogue of 'Alfin' polybutadiene* and had painstakingly purified a quantity of isoprene for that purpose.

(2) The Goodrich spectroscopists were exceptionally well equipped to recognise the infrared 'fingerprints' of an authentic *cis*-1,4-polyisoprene. They had been called upon to provide a sure-fire identification of natural rubber in the sealing mixtures used in the original tubeless tyres (in connection with patent infringement suits), and as a result of this experience the infrared spectrum of natural rubber was as familiar as the face of a friend.

(3) Horne was following Ziegler's polyethylene 'recipe', not trying to duplicate his earlier work on polymerising diolefins; hence, he did not fall into the error of using ether as a reaction solvent. Moreover, the inert hydrocarbon solvent he did use kept him out of the difficulties of bulk polymerisation which were plaguing certain other workers (see below).

---

*Butadiene polymerised with a complex organosodium catalyst, developed by Prof. Avery Morton at M.I.T. It gave a linear polymer of tremendously high molecular weight, but not stereoregular.

(4) Perhaps most importantly, he did not allow himself to be deterred by the indications from the 'Master' (Ziegler) that diolefins would not polymerise satisfactorily.

(5) He was lucky. His first attempt with isoprene succeeded; the next five failed, for one reason or another. Had it been the first five that failed, it is doubtful that there would have been a sixth. 'Old research hands' will recognise this as a recurrent pattern in the history of discovery: it is surprising how often the first experiment has given the good and novel result which in several succeeding attempts cannot be duplicated. Logic argues that there must be innumerable other discoveries yet to be made—missed in the past because some first attempt failed and was not repeated.

(6) Finally, his company was in an expansion phase and management in a correspondingly aggressive mood.

As a result of the latter circumstances, the laboratory became a place of excitement and hard-driving effort that involved keeping the work going around the clock and on weekends. So many people from Goodrich and Goodrich-Gulf, up to and including the Chairman of the Board, visited Horne's laboratory to witness his 'show and tell' that he felt he should have had bleacher seats installed.

The patent attorneys also descended on him and his co-workers to prepare the patent application, and when he signed the paper assigning rights to Goodrich-Gulf, he was given the then-customary one dollar (the 'good and valuable consideration received' acknowledged in the assignment*) with more ceremony than usual. There was just one slight hitch. This was the first patent assigned to the joint company, Goodrich-Gulf, and no provision had been made for funding such events. The problem was solved when one of the lawyers put up the dollar from his own pocket. This bill was then initialled by the Chairman of the Board and other company officers, one of whom drew a 'smiling face' on it. Needless to say, that particular dollar has never been spent.

There was not much else in the way of celebration and ceremony, but there was continued serious and intense effort. The laboratory program was reorganised for maximum effort on completing the process research and engineering, product evaluation and scale-up to pilot plant. Budgets were forgotten; anything needed by the chemists was ordered without question, for delivery by air. Strenuous attempts were made to

---

*Better and more valuable considerations were forthcoming later, but at Goodrich's option.

maintain secrecy. The requirements for large amounts of isoprene monomer, then only a development chemical, was not only a supply problem but a potential giveaway. It was handled by having numerous small purchases made from different places using a variety of names.

This and other security measures seemed to have worked fairly well; although some rumours seeped out, they did not have the force and conviction to tip off the trade press or the competition. The best protection was the fast pace of the program. One of Horne's laboratory assistants, a young girl technician, was not even aware that she was involved in something really big for almost three weeks. She then berated him for not giving her the inside information in the beginning.

As soon as convincing data were on hand and patent applications in condition for filing, Goodrich-Gulf management were ready to act—only a few months after the initial discovery. The managers of Goodrich plants all over the country were called in for a meeting at which they were given the whole story in confidence; four patent attorneys were then put on a plane to Washington to file the application in person, and the public announcement was readied for release the following day.

This expeditious and aggressive policy gave Goodrich and Goodrich-Gulf the credit, the publicity* and the competitive edge in what the Akron Beacon-Journal said might rank as 'one of the most important scientific achievements of the century'. It stands in striking contrast to the caution and delays exercised by their competitor, Firestone, where synthetic polyisoprene had actually been made much earlier but had been kept under wraps (see below). Goodrich-Gulf went ahead to build a large pilot plant at Avon Lake, Ohio, followed by full-scale manufacture at Institute, West Virginia. The time from the first laboratory experiment to commercial production was less than five years—under half the time usually required in the chemical industry.

---

*But publicity is as perishable as fame is fleeting. In 1973, an article in *Chemical and Engineering News* gave credit for the discovery of *cis*-1,4-polyisoprene to Good*year*. A correction was promptly, if somewhat wearily, issued by Goodrich, who are so accustomed to being victimised by this confusion of names ('The Curse of the Goodrich Name') that they have even used it as the basis for an advertising compaign ('We're the *other* guys!').

But what about the 'pool'?

As related above, the Goodrich story is a generally admirable saga of smart, opportunistic research and aggressive, alert management. But the bright shield is not without tarnish. Goodrich's discovery announcement immediately raised eyebrows and questions among the other companies who shared, with Goodrich, membership in the synthetic rubber 'Pool' agreement. Since that agreement called for exchange (by publication in *'Rubber Reserve Reports'* circulated to all members) of all information in the field of 'synthetic rubber'—defined as any rubbery polymer containing at least 50% of any diolefin—how could Goodrich do research on synthetic polyisoprene and withhold it from the *Rubber Reserve Reports*? Moreover, were not the Government and all pool members automatically entitled to royalty-free licences and the accompanying know-how by the terms of the agreement?

These questions were raised officially by attorneys for Rubber Reserve. Goodrich-Gulf answered, and repeated later in public, that their discovery had been made with the use of catalysts and information purchased from Karl Ziegler and that the terms of their licence from him required that they not disclose any of Ziegler's information or grant any rights to third parties. They also drew a distinction between B. F. Goodrich, the party to the Rubber Reserve pool, and Goodrich-Gulf, who held the Ziegler licence and had 'assigned' the research team to the B.F.G. laboratory.

These arguments carried little conviction with the rest of the industry and none at all with Rubber Reserve. Goodrich-Gulf's own lawyers had recognised from the beginning that they had a problem, for at one point they had advised Horne: 'You had better put some ethylene in it [his polyisoprene system], even though you don't work for Goodrich. They might construe Goodrich-Gulf as being under the Rubber Reserve licence'.

Over the next few years, negotiation was succeeded by litigation, and eventually Goodrich accepted a settlement that required them to license others.

The Firestone story ('another Firestone first' comes second)

Chronologically, Firestone was first with polyisoprene and polybutadiene, for their chemists had made 'high-*cis*' polymers before Horne, but because of management indecision and delays, they were 'scooped' by Goodrich's announcement and relegated to a poor second in public awareness and recognition.

In common with the other rubber companies in the U.S., Firestone's polymerisation research in the post-war decade was concentrated on improving the standard system of polymerisation in aqueous emulsion with a free-radical catalyst—a legacy from the pre-war German 'Buna S' and the American 'GR-S' and 'SBR' (all butadiene–styrene co-polymers). Sooner than most, Firestone became disenchanted with this area and decided to seek greener fields for research. This decision was made by Dr Fred W. Stavely, Firestone's Director of Research, who also played an unusually direct role (for a research director) in the discoveries that ensued. He convened his key research staff in the summer of 1952 to express his dim view of the prospects for real advances if they stuck with the conventional systems, and to solicit ideas for new directions, particularly those that could lead to high-*cis*-content polymers more nearly resembling natural rubber.

Fred Stavely may fairly be described as an earlier-generation leader and elder statesman in the U.S. rubber industry. Born in the last century, he took a Master's degree in chemistry at the University of Chicago in 1917. What makes that event noteworthy is the title of his thesis: '*Synthetic Rubber*'! More noteworthy still are two conclusions he drew from that literature survey.

(1) 'To make synthetic rubber a success, the crude product would have to be placed on the market at 24¢ per pound'. Twenty-five years later, the price set (and maintained for years) for the U.S.-made synthetic, GR-S, was 23¢ per pound!

(2) 'We have therefore before us in the investigation of synthetic rubber alone, a field of chemistry that cannot be surpassed in interest, has great practical possibilities and, when completely solved, will be of marked importance in the field of pure chemistry'.

Not surprisingly, this too-early prophet went without early honours. After three years on a Sumatra rubber plantation, Stavely returned to the U.S. jobless, went back to Chicago, and earned a Ph.D. in organic chemistry in 1922 under Prof. Kharasch. Those interested in identifying the spawning grounds for scientists, particularly polymer scientists, will be impressed by the fact, too remarkable to be a coincidence, that no less than four men who became research directors or technical directors for the 'Big Four' rubber companies got their graduate training within a few years of each other in the same Chemistry Department at Chicago*.

Stavely started his career at Firestone in 1922 as a research chemist. He can therefore be included among the 'pioneers' of rubber research in the U.S.; but that classification is misleading, since he soon moved into technical management and then through an unusually varied series of positions in sales, personnel and manufacturing, before coming back into research management after World War II. He was not altogether a 'research man's research director', because his varied experience gave him some outlooks that were unsettling to the pure research viewpoint; on the other hand, he displayed a determination to try new approaches and to back his ideas and his men that was reassuring to them and unsettling to his superiors. The remark by one of the men who worked for him is revealing.

'For some reason', he told me, 'Stavely was always just encouraged by any result, that was different. He was always bullish'.

The man who made that comment, Fred Foster, was one of the research chemists who attended Stavely's 1952 meeting mentioned above. Foster had come to Firestone in 1946 after getting his Ph.D. in physical chemistry from Columbia and two years of experience in Hercules Powder Co. His initial assignment, as Sam Horne's was to be a few years later at Goodrich, was to read up on the work of Mayo, Walling and Lewis on co-polymerisation ratios and then to extend that work to cover monomer pairs of interest to Firestone. He began 'grinding out' such measurements in the standard, radical-catalysed emulsion system and then, in order to 'add a little spice', turned to ionic catalysts, beginning with metallic sodium.

In the literature on this field he found the early papers of Karl Ziegler on polymerisation of butadiene and Ziegler's prediction (made in 1937!) that high-*cis* polydienes should be attainable through sodium catalysis.

In response to Stavely's solicitation, Foster, who is characteristically cooperative, outgoing and even voluble, did his best to make a positive response. In a memorandum of July 11, 1952, he reviewed the discouraging prospects for achieving the desired 'microstructure' with radical catalysts or by varying the reaction conditions, but pointed to a report of Morton at M.I.T. describing a sodium-catalysed polybutadiene with higher *cis* than *trans* structure. He also cited Ziegler's 1937 prediction of the attainability of high-*cis* polydienes. However, since nothing even

---

*The others: Harlan Trumbull (Goodrich), Sydney Cadwell (U.S. Rubber), and Herman Thies (Goodyear).

remotely approaching an all-*cis* synthetic polymer had been made or seemed likely to be, Forster could not generate any real optimism or advance any positive recommendations.

The striking thing about the two observations cited by Foster is that both were erroneous. The M.I.T. structural analyses were later shown to be seriously in error (first pointed out by Prof. Maurice Morton—no relation to Avery Morton—at the University of Akron), and so was Ziegler's expectation about sodium (at least, so far). Fortunately, both errors were on the optimistic side; otherwise, there would have been no mention of them and no grounds, even tenuous, for encouragement. As it was, however, Stavely found Foster's memorandum interesting and encouraged him to continue investigating ionic polymerisations.

Foster had also been encouraged and obtained a prophetic comment from a source he greatly respected outside the company—Paul Flory, whom we met earlier (Ch. 2), and who was presiding at a meeting at which Foster gave a paper (he was allowed to publish some of his earliest results on polymerisation with sodium because they met the standard criteria for publication by an industrial research chemist: very interesting to scientists but not to the company that paid for the work). Flory found the data of sufficient interest that he sent word to Foster that he ought to continue working with ionic catalysts because: 'You might come up with something interesting'.

As Foster continued working and reading, he encountered Ziegler's papers on organic compounds of lithium and wondered why no-one had pursued lithium as a polymerisation catalyst*. He decided to try lithium but, being familiar (obsessed?) with catalysts in the metallic state, he used the metal instead of the organic lithium compound that might have saved a lot of time and trouble. In a co-polymerisation system he got enhanced *cis* content in the diene portion—35%. While this was twice as high as in the standard emulsion system, it was such a far cry from the desired 95–100%, that Foster felt it was of little significance and put the data in his monthly report quite matter-of-factly.

He was surprised, therefore, then Stavely called him in to talk about the report. It was no surprise that he had read the report, for it was

---

*In a 1929 paper, Ziegler had mentioned polymerisation of butadiene with a lithium compound, but had obtained only slow reaction and low-molecular-weight polymer. He doubtless was handicapped, as noted above, by having to work with relatively impure butadiene and by the fact that he used ether as a solvent, which is now known to have a deactivating effect on the catalyst.

well known that the research director somehow managed to read all the reports in the whole laboratory, but his interest in this particular result was unexpected.

'Fred', said Stavely, 'you're getting an unusual structure there'.

'Yeah', said Foster, 'it's different, but it's not too encouraging.'

'Well', Stavely persisted, 'it is different, though. Don't you think we ought to try that also on isoprene?'

Foster had no real enthusiasm for that idea, but it is always a good idea to agree that the boss's idea is a good one, so he did. Actually, Foster thinks Stavely probably already had lithium in mind from his own reading of *Rubber Reserve Reports* and the literature.

The very next day a co-worker, Lynn Wakefield, came to Foster and asked to borrow some lithium from him. A quiet, even taciturn worker, Wakefield did not volunteer his reason, but Foster immediately guessed it, and Wakefield acknowledged that Stavely had asked him to try it on isoprene. Foster had no objection to letting someone else try this 'way out' experiment and readily supplied the sample of lithium.

Wakefield's initial attempt with isoprene was too successful; the reaction was so violent that it blew up his reaction vessel. He did salvage some charred polymer from the wreckage and, proceeding more cautiously in subsequent tests, eventually managed to get something that looked like rubber.

He now faced another problem—how to find out what kind of polymer he had. At that time, Firestone's spectroscopic expert, John Binder, had good procedures for analysing polybutadiene by infrared spectroscopy, but not for polyisoprene (Goodrich's capability was probably just being developed, and of course was not published until much later). X-ray diffraction measurements would not be of use until a sufficiently high *cis*-1, 4-content (over 90%) had been achieved to induce significant crystallisability. So, lacking elegant analytical tools, Wakefield did what an experienced rubber chemist would do—he scraped together enough of a sample to compound and vulcanise it (i.e. he mixed it with sulphur and accelerators and cured it under heat and pressure), making a specimen that could be subjected to elementary physical testing.

Stavely presided at another meeting at which Wakefield's very limited test data on his crude sample were evaluated. The chief compounder for Firestone, invited to give his assessment, said:

'Those properties certainly aren't very impressive.'

'No', replied Stavely, 'but look at the gum tensile' [tensile strength of sample containing no reinforcing agent—notoriously poor for synthetics].

'It's nothing compared with natural rubber', objected the compounder.

'No', agreed Stavely, 'but it's still high compared to any emulsion polymer.'

Acting once again on maximum encouragement derived from minimum data, Stavely launched an intensified program that moved the project into pilot plant in scaled-up equipment in September 1953. Working now in steel reactors behind sandbagged barricades, the team contended with explosions, fires and plugged-up equipment. Since the system had no diluent, a successful reaction (one that neither died, blew up nor blew out) produced a solid block of rubber inside the vessel.

'Coral Rubber'

Once, when a reactor was allowed to stand over a weekend before being opened, the investigators were surprised to find that some of the polymer had grown in the upper part of the reactor as clear, water-white nodules resembling coral in structure. When plunged into a solution of stabiliser (to prevent oxidation), it took on an orange colour that heightened the resemblance. Fred Stavely, on first seeing this sample, christened it 'Coral Rubber', and the name stuck.

This clear rubber was tested separately and was found to have a high enough *cis*-1,4- content to show an x-ray crystal pattern when stretched! Now, at last, there were solid grounds for both scientific and commercial optimism, for it could be said with confidence that the first synthesis of a polyisoprene approaching natural rubber in composition and properties had been achieved.

Efforts were redoubled to make reproducible samples of high-*cis* polymer by a reproducible process; concurrently, one might have expected that Firestone management would have swung into action with a public announcement of the technical achievement and plans for commercial production. They did neither, for the very good reason that top management were not even aware of what had been going on!

For reasons best known to himself, John Street, Firestone's Vice President and Technical Director, and Stavely's immediate superior, was not convinced that Stavely had brought him a real 'winner' (no doubt, there had been a few too many cries of 'Gold, gold!' before), and was not even willing to put the whole story before the president and the board. Stavely, therefore, could not get authorisation for the stepped-up and scaled-up development effort he knew was now essential.

'So I bootlegged the money for it', he says. 'If they had known all I was doing, I'd have been fired'.

The problem, in Stavely's opinion, was that top management could not appreciate the significance of an x-ray diffraction diagram and would not make commitments on the strength of such esoteric evidence, backed only by laboratory test data. Street probably shared the deeply felt and oft-voiced tradition of the tyre industry: 'Unless it's been tested in tyres, you don't know if you've got anything.' So he insisted on waiting until there was enough rubber to make tyres and the tyres had been road tested. This, of course, took many months.

When Goodrich dropped their bombshell, announcing the laboratory synthesis of a duplicate of natural rubber, Firestone already had been in pilot plant, however covertly, for over a year, and had tyres of synthetic polyisoprene on the road. Firestone's president, Harvey Firestone, Jr, and other top management were hastily informed fully about what their own research had accomplished, but even then they did not choose to make any public announcement until another half-year had passed. ('Much to my disgust!', says Stavely.) Finally, in August 1955, a press release announced that Firestone Research had synthesised *cis*-1,4-polyisoprene by a proprietary process. By that time, such news had a definite 'me, too' flavour; nevertheless, it was greeted with strong interest in the industry because of what Firestone said about licensing possibilities.

Unlike Goodrich, Firestone had concluded early in the game that their work would come into the field of the Rubber Reserve agreement and would be subject to its licensing provisions. Goodrich's position that their discovery fell outside the 'pool' area was 'absolutely preposterous' to Firestone personnel. This attitude reflected a difference in situation more than in saintliness. Nobody is perfect, and no corporation is more nearly perfect than it needs to be.

Although their plans to investigate lithium had been mentioned in the *Rubber Reserve Reports,* there was a substantial delay in reporting the exciting results obtained with it. Fred Foster's work was not budgeted or reported under the Rubber Reserve contract, but Wakefield's was; nevertheless, when he made his startling polyisoprene discovery some of his reports were hastily diverted and kept out of the reporting system for a time. Careful re-reading of the Rubber Reserve contract, however, convinced the Firestone lawyers that they had no chance of maintaining a proprietary and exclusive position. Unlike Goodrich, they had neither a separately owned company nor an outside licencer with which to confuse the issue.

So Firestone indicated from the beginning that licences would be available, and interested parties were soon standing on Firestone's 'Welcome' mat. I was one of these, since I had friends in Firestone and my company also had an active research program in the same field. At this meeting, Firestone research people, in the presence of their attorneys, described the bulk polymerisation (without naming the catalyst), as one which was conducted in glass 'pop' bottles and which was terminated by cutting the bottle in half to remove the block of rubber.

'And just how', I asked, 'will this work on commercial scale? Do you visualise cutting a 5000-gallon glass reactor in half to remove a 16 ton block of rubber?'

They were not particularly amused.

'We have other ways', was all they would say.

Actually, they probably did not, in terms of processes proven at that time, but two important improvements were in the works. Larry Forman, another Firestone research chemist, had tried in vain to convince Foster that the reaction could be controlled better if run in solution, rather than bulk. Foster's objections were logical: solvents were very apt to retard polymerisation excessively, cut the molecular weight of the polymer, and possibly interact with the catalyst (compare this with Ziegler's experience with ether). But Forman, who describes this as a 'friendly controversy', persisted in his view and undertook the trials himself. Just as Foster had predicted, the reaction became too slow to be practical. But Forman counteracted this effect by putting roller bars inside the rotating reactor, thus keeping the surface of the catalytic metal particles bright and achieving good reaction rates, high molecular weights and high conversions without solidification.

The final major improvement was made by still another chemist, Dick Stearns, who proposed and demonstrated that an organic compound of lithium—a lithium alkyl—is a soluble catalyst that gives the desired reaction rates and conversions while eliminating mechanical problems associated with a dispersed metal catalyst. Thus the circle was closed by coming around at length to the organolithium compounds first described by Karl Ziegler in the 1930s that, in hindsight's bright afterglow, would appear to have been a logical starting point.

We may pause here to note that Fred Stavely's name has figured in the Firestone story with unusual prominence for a research director, whose concerns with budgets, personnel and keeping management happy would normally preclude his participation in the research itself. But, as we have seen, Stavely was somewhat like Karl Ziegler in his insistence on direct involvement and in his willingness to risk a little management unhappiness in order to advance the research. He was also willing to risk offending his subordinates when he bypassed them to deal directly with the chemist at the bench. At times, he would call in Fred Foster, for example, for a private discussion without going through Ray Dunbrook, his Assistant Director, or Ben Johnson, Foster's immediate supervisor. This was so habitual with him that at a round-table conference at the Industrial Research Institute on 'The First Level of Supervision', Stavely's own description of how he operated elicited the comment from another director: 'Fred, at Firestone, *you* are the first level of supervision!'.

However much Stavely's 'style' may have offended principles of management and some of his people, it was undoubtedly his personal involvement and persistent pushing that brought about the discoveries and the developments at Firestone. The chemists who were the actual inventors and whose names appear on the patents freely acknowledge that it was Fred Stavely who advanced the original idea and who stimulated, prodded and encouraged them into making it a reality. In rather belated recognition of his key role, the Rubber Division of the American Chemical Society awarded him its Charles Goodyear Medal in 1972.

Goodyear: the last shall be first (to publish)

Goodyear, biggest of the Big Four, apparently had neither program nor premonition about the impending duplication of natural rubber, and so was caught flat-footed (flat-tyred?) by Goodrich's announcement. Fortunately for them, however, one of Akron's strongest taboos was violated at a cocktail party given by President Thomas of Goodyear.

Even more strongly than in most one-industry towns, the First Commandment in Akron has always been 'Thou Shalt Not Discuss Company Affairs At Social Affairs Attended By Competitiors'. As with the original Commandments, however, top executives sometimes consider themselves above the rules made for the common man. At Thomas's Sunday afternoon cocktail party, he was talking over a drink with

J. E. Trainer, an executive vice president of Firestone, and said to him, in effect: 'You know, Goodrich have come out with this statement that they have synthesised natural rubber. That sort of puts us in a bind, doesn't it? What are *we* going to do?'

Whether spontaneous or subtle, Thomas's approach to disarming the opposition was successful. Trainer may have been caught off guard by the conspiratorial 'we', flung with unusual familiarity around his shoulders. Having just been briefed on what his own research group had accomplished, Trainer could not resist boasting that Firestone were not worried, because they had accomplished the same feat as Goodrich much earlier and had a process of their own under development.

The next morning, Dr Osterhof, director of research at Goodyear, telephoned Dr Raymond Dunbrook, second in command at Firestone Research under Stavely, to ask: 'What's this that Trainer told our president yesterday–that you guys have a synthesis for *cis*-polyisoprene? Maybe we should get together!'.

Eventually they did and, within a few months, an information exchange agreement had been signed and the Goodyear and Firestone research departments were having joint meetings. What Goodyear offered to put into the pot was their process information, then quite incomplete, on synthesis and purification of isoprene monomer.

This agreement, shortlived as it turned out to be, was remarkable more as an unusual, if not unprecedented, cooperation between two traditional rivals than for the results it produced. Firestone participants felt that Goodyear had relatively little to offer, and Goodyear evidently felt that the Firestone product and process were not fully competitive with Goodrich's. The exchange agreement was soon allowed to lapse, by mutual consent. Goodyear eventually got a Ziegler licence and went into commercial production with their own variant of that process. They also used a process for making isoprene monomer that started with propylene dimer made by another Ziegler catalyst, so Ziegler is unquestionably the main father (and a major beneficiary) of the commercial manufacture of polyisoprene in the U.S.

The Shell story: first in, first out

While all the things related above were going on among the rubber companies in Akron, some of the big oil and chemical companies were also trying to get into the act. They saw an enormous potential market for petroleum-derived monomers– but they also saw more potential profit if they could convert their monomers into polymers (rubber) themselves. To the rubber companies, on the other hand, the spectre of the major oil companies becoming integrated from well to rubber, perhaps to tyres, had long been a consummation devoutly to be resisted, even back in wartime. They had made sure that the oil companies were not included in the government-sponsored research 'pool' and Rubber Reserve meetings, and that they were strongly encouraged to be content with, or at least resigned to, being raw material suppliers only. The extensive cooperative research and production programmes between the oil and rubber companies therefore also had competitive aspects, and it was sometimes hard to tell which was dominant.

The rubber companies' strategy was successful for the duration, but when the government-owned synthetic rubber plants that made butadiene, styrene and the co-polymer rubber, GR-S*, were auctioned off after the war, the oil companies finally got their chance. Thus Shell Chemical, for example, entered the rubber business in 1955 with the purchase of the West Coast (Torrance, California) plants that made butadiene, styrene and co-polymer. Having been a pioneer in butadiene and an operator of the diene plant, Shell had a long-standing interest in butadiene and in expanding its uses. That goal was pursued in a substantial research program in the Shell Development Company laboratories at Emeryville, California, aimed at both chemical derivatives and polymers. Dr Lee Porter, a chemist engaged in such research, concluded, after perusing the literature, that lithium was the most promising catalyst and that a soluble form would have practical advantages.

Porter's first experiments were promising and his later ones highly successful. Numerous co-workers then joined him in developing a process which soon was capable of turning out polybutadiene of high steric purity. Enthusiasm mounted quickly, because management saw an outlet for butadiene and the chemists saw a chance to make good on their boast that synthetic rubbers would eventually surpass natural rubber in overall quality.

Cis-1,4-polybutadiene is, indeed, even more 'rubbery' than nature's rubber or synthetic polyisoprene. It has more 'bounce' (witness 'Superball') and makes tyres that run cooler and wear longer. Also, it stiffens very little at low temperatures and resists 'taking a set'. These seemed, and were, important advantages. True, they were accompanied by

*For Government Rubber–Styrene (type); later rechristened SBR, for Styrene–Butadiene Rubber.

what seemed like minor disadvantages; particularly, a distressing tendency to skid on wet pavements. Also, small cuts in tyre treads tended to grow into big ones.

Research, typically, was optimistic that these difficulties could be overcome, because they were the sort that should yield to compounding variations. But a great deal of work was done toward that end with no significant progress. Meanwhile, other Shell scientists had worked out a process for making isoprene monomer from refinery gases and had found modified polymerisation catalysts that made a fairly high-*cis*-content polyisoprene. Shell management, who were painfully aware that for an oil company to sell rubber to rubber companies was about as easy as to sell oil to Arabs, concluded that it might be slightly less difficult if they were to offer a match for the known product, natural rubber, instead of the unproven polybutadiene rubber. Perhaps this was another case of marketing's preference for the known over the unknown, as noted in Ch. 6. In any event, the process development and plant design teams were swung over into an all-out effort to get into the earliest possible production of synthetic polyisoprene.

The rubber companies, who had first discovered *cis*-polyisoprene and had loudly hailed its significance, delayed their own commitments long enough that Shell was able to become the first company to bring a commercial plant on-stream. But the triumph was shortlived; Shell was also the first to get out. Their polyisoprene had turned out also to have some disadvantages; it was not quite equal to natural rubber in *cis*-1,4-content and this barely measurable difference proved to have a quite measurable effect on product performance. Short of switching to an entirely different catalyst system, which would have entailed licensing problems, no means were found to bring the *cis*-1,4- content fully up to the natural rubber level. And so, after a dozen years of competing with natural rubber and with the 'captive' production of synthetic polyisoprene by the rubber companies for their own needs, Shell Chemical abandoned polyisoprene production and converted their plant to a still newer, and unique, speciality synthetic rubber.

The A.C.S. meeting

As far as the scientific community was concerned, all the preceding ferment came to a head in the autumn of 1955 at the meeting of the Rubber Division of the American Chemical Society in Philadelphia (November 2–4). Goodrich and Firestone both were listed on the program with papers on synthetic polyisoprene. R.P. Dinsmore, Goodyear's Vice President and Technical Director, was scheduled to receive the Goodyear Medal, an annual award passed around by the Rubber Division to honour the rubber establishment's recognised pioneers*.

Through the program announcement and advance abstracts of papers, Goodyear knew of their competitors' papers and that they would be 'show-stoppers', whereas their own research had been 'too little and too late' to qualify an entry of their own. This was not a happy position for either Goodyear or the man who was to be honoured at the meeting. Goodyear therefore did what they could with the results of a hastily mounted crash program in which they had used a Ziegler-type catalyst to make *cis*-polyisoprene in the laboratory, at least in sufficient quantity to get an infrared spectrum. They moved to 'scoop' the A.C.S. meeting and their competitors' publications by issuing an announcement on the very eve of that meeting. Dr Dinsmore was quoted as saying:

'Our research scientists have found that a heterogeneous catalyst system of the so-called Ziegler type . . . is also effective in . . . polymerisation of isoprene. . . . The polymer is nearly all *cis*-1, 4-.' He also stated that there was a 'vast difference between our catalyst system and Firestone's'. Remarkable by its absence was any comparison with, or even acknowledgement of, the existence of Goodrich's catalyst system and polymer. The real item of novelty and interest was that the press release included graphs that reproduced tracings of the infrared spectra of the synthetic polyisoprene and natural rubber.

Whether by good timing or by pulling strings and pushing people, Goodyear managed to get this announcement printed in the October issue of the trade publication '*Rubber Age*' and in the October 24 issue of '*Chemical and Engineering News*'. The papers presented the following week at the A.C.S. meeting did not appear in print until many months later; thus Goodyear became the first to publish a comparison of spectra as evidence of stereoregularity in polyisoprene.

The spectacle of the three big rubber companies manoeuvring for credit, priorities and publicity, and the prospect of seeing the inventors and learning something definite about their inventions, generated a higher level of attendance and attentiveness than most A.C.S. sessions enjoy. But the audience's expectations were only partially fulfilled;

---

*The medal is not from the Goodyear Rubber Company; both the company and the medal are named after Charles Goodyear, the inventor of vulcanisation.

the papers were given, not by research chemists but by research executives, and only one offered any data at all on the polymerisation process used.

The Goodrich paper, given by Frank Schoenfeld, Vice President for Research, discussed at length the properties and performance of the 'synthetic natural rubber', Ameripol SN, but gave not a hint as to how it was made. This displeased the technical audience, as did Schoenfeld's refusal to answer questions on the grounds that the terms of Goodrich's Ziegler licence would not permit it.

Fred Stavely, who presented Firestone's paper on 'Coral Rubber', was the beneficiary of the situation. His smooth presentation included disclosure of the catalyst and conditions used—in fact the exact polymerisation 'recipe' and conditions were revealed. Probably for the first time in the history of Chemical Society meetings, the simple recitation of the experimental conditions was greeted with loud cheers, and when Stavely finished he was given a standing ovation. The audience was expressing its feelings against Schoenfeld as much as its appreciation of Stavely.

The final outcome

With that recognition and accolade, the optimism displayed by Stavely and his research staff was finally justified and their persistent efforts crowned with success—almost. The qualification is necessitated by a seemingly trivial difference that proved in the long run to be both significant and unsurmountable. Coral Rubber had, at best, a *cis*-1,4-content of about 94% compared to 98% for natural rubber. Research people at Firestone took the attitude that the difference probably was not important but, if necessary, it could be eliminated by further research and development work. But experience showed that there were at least minor disadvantages associated with even a few per cent lower *cis* content, and Firestone had no more success than Shell in closing the gap.

At Firestone, Larry Forman had once boasted that Goodrich's obstreperousness about licensing need concern no-one, because no-one would ever use the process anyway, given the availability and advantages of Firestone's process. Ironically, it was the Firestone process that was never used for polyisoprene. Firestone management, weighing the same technical factors as Shell, came to just the opposite decision: they opted to shelve polyisoprene in favour of polybutadiene, which they felt offered fewer problems.

Offhand, it would appear that one of these decisions had to be wrong; actually, however, both may have been right—or both wrong. The two companies faced entirely different marketing situations: Firestone was concerned primarily with satisfying the needs of their own tyre builders (not that that made it easy, as anyone who has tried to 'sell' to presumed 'captive' outlets in his own company can testify), whereas Shell was the outsider trying to break in.

Another vote for polybutadiene over polyisoprene was cast by Phillips Petroleum Company. In their laboratories, birthplace of the most successful linear polyethylene process (Ch. 5), another of the numerous 'close second' discoveries that have marked our story was scored by Dr R. P. Zelinsky. By his own account, Zelinsky discovered *cis*-polyisoprene independently, but just a week later Goodrich's public announcement made it evident that he had not been the first. And in patent, as in political, races, there is no prize for second. Later, however, Zelinsky discovered, and was able to patent, a separate process for making *cis*-polybutadiene. With characteristic aggressiveness, Phillips launched a manufacturing and marketing program on that polymer with the same energy they had displayed earlier with carbon black and with polyethylene, and with at least modest success.

Despite these and other major commercial commitments to polybutadiene, however, that polymer has never fulfilled the early hopes held for it as the 'rubberiest rubber'. Because of secondary performance problems that have still not been fully overcome, it is still used almost exclusively in blends with other rubbers. Even so, its total usage is several times as great as that of synthetic polyisoprene, and the final chapters on its development remain to be written.

As for the polyisoprene story, it ends with the '*Good-*' guys—Goodrich and Goodyear—winning both game and 'rubber', as the only remaining producers in the U.S. And both use a Ziegler-type catalyst under a Ziegler licence.

11 The Crest of the Wave — and the Trough

If I had all truth in my hands, I would be careful not to open them.
–Fontenelle

In the first decade after their discovery, the stereoregular polymers were hailed as materials of great promise, and in the second decade they fulfilled that promise. In this narrative, detailed statistics are out of place and would be out of date when printed; however, we may at least note that by 1975 over sixty plants had been built by some fifty different companies to produce such polymers in a dozen different countries. Today the polyolefins (including the original member, high-pressure polyethylene) have achieved dominance among synthetic plastics. Their production volume is far greater than that of any other class, and the total even exceeds the combined total of all other plastics.

What, then, of all the others?

That being the case, it is obvious that the people, laboratories and companies that have appeared in these pages are but part of the cast, and the period described thus far is but the beginning, of the saga of the stereoregular polymers. Our concern, as stated in the introduction, is primarily with beginnings: the key discoveries and the interactions of people and events that led from discovery to fruitful innovation.

Nevertheless, there are individuals and organisations who can rightly claim to have made significant contributions at early stages and to deserve mention as much as some of those that have been discussed. A few U.S. companies have received the lion's share of attention, but many others also became producers of stereoregular plastics or rubbers and contributed to the growth of technology and the patent literature.

Still other companies figured importantly in developing catalyst components and supplying them to industry. And many scientists, unnamed here, have worked diligently in university and industrial laboratories to provide understanding of catalyst structure and reaction mechanisms.

But this is neither a scientific treatise nor a history of commercial development. It is a story about certain people, certain seminal events which they brought about and selected subsequent happenings that round out the story and point up its significance. These happenings illustrate the interactions that must occur and the hits, misses and setbacks that inevitably accompany them in the evolutionary course that a research finding takes on its way to becoming a commercial success.

Many another 'case history' could be adduced to extend the account of middle- and late-stage developments, but at the risk of becoming tediously repetitive. Some pieces that might have enlightened our narrative are missing because a number of industrial laboratories that did good work did deplorably little publishing, or published so late that the timely significance of their work was lost*.

Other omissions doubtless occurred because the author was not sufficiently diligent or was not clever enough to know whom to ask what, and when. In any event, the conclusions drawn must rest on the facts at hand, and their validity on the interpretations made from them.

Impact of linear and stereoregular polymers on industry

Although corporate planners like to assure management that they are the captains of their fates, in actual fact the destiny that shapes their

---

*It is a source of regret and frustration to one attempting to put together a story that is true to the facts and fair to the people involved that so few of those people chose, or were allowed, to publish accounts of their discoveries. In striking contrast to the prolificacy of Ziegler and Natta, most of the others who contributed to this huge field have published relatively little and late. This is doubtless because the great majority are in industrial laboratories. Many companies whom one might expect to be proud of their research accomplishments have not only refused to publish them but to this day will not even acknowledge, much less accept, invitations to supplement the incomplete public record with statements of fact or position.

A reasonable caution on the part of those still engaged in patent litigation is understandable, but caution should stop short of paranoia. Those who hide their heads in the sand have little grounds for complaint about how others describe what remains visible.

corporate ends is often new technology whose advent and impact cannot be foreseen, but which creates opportunities for the opportunistic. As we have seen, the major companies that undertook production of stereoregular polymers found it expedient to form new alliances, subsidiaries and joint ventures for that purpose, accompanied by a larger number of second-derivative fabricating operations to convert the polymers into fibres, film, packages, moulded articles and other finished products. Many of these ventures were successful; quite a few were evanescent or outright failures. Some integrated marriages between U.S. corporations were scarcely consummated before they were annulled as

TABLE 11.1 *U.S. companies who are, or have been, producers of linear and stereoregular polymers.*

Linear Polyethylene

Allied Chemical
Amoco Chemical
Celanese
Chemplex
duPont
Exxon (Standard Oil Co.(N.J.))
Grace
Gulf
Hercules
Monsanto
National Distillers
National Petrochemicals
Phillips
Sinclair–Koppers
Union Carbide

Synthetic Rubber

American Chemical & Rubber
Copolymer Corp.
duPont
Exxon
Firestone
Goodrich-Gulf (Ameripol)
Goodyear
Phillips
Shell
Uniroyal

Polypropylene

Alamo Chemical
Amoco Chemical
Arco Polymer
Avisun
Dart Industries
Diamond Shamrock
Dow Chemical
Eastman
Exxon
Gulf
Hercules
Northern Petrochemicals
Novamont
Phillips
Rexall (Rexene)
Shell
Soltex

miscegenations on anti-trust grounds, and their fabricating subsidiaries declared illegitimate offspring.

By what is essentially a trial-and-error approach, certain technologies, products and business entities have emerged as the fittest to survive, but they have no guarantee of continued viability. In fact, any detailed description of current combinations would be of little value, since present patterns are bound to be altered in the never-ending course of technical and economic evolution.

Nevertheless, some overall indication should be given of the total industrial involvement. For the U.S., table 11.1 provides (hopefully, without serious omissions) the names of the companies that are, or have been, producers of stereoregular polymers. The enumeration is in alphabetical order, and no chronology or ranking is implied. There are more names than active producers, since some companies (or plants) have been bought by or merged with others, and some company names have changed.

Impact on the public

As these polyolefins and polydienes move from the laboratory to the plants that manufacture them, thence to the plants that make useful products from them, and finally on to us, the people who use those products, there is a tremendous escalation at each step in the numbers of people and locations involved. The polymer manufacturing plants are numbered in the tens, the converting (product manufacturing) plants in the thousands, and the users in the hundreds of millions. The applications whose trial-and-error beginnings were described in previous chapters have proliferated to the point where any attempt to enumerate all the ways in which the products now enter into everyday life would prove an endless task.

In the form of fibres (carpeting, upholstery, sacks, for example), film (bags, wrapping film), rubber goods (tyres, sporting goods), blow-moulded containers (for milk, detergents, cosmetics), uncountable other moulded articles (houseware, automotive parts, toys, appliances, seating, etc.) and innumerable other items and shapes, stereoregular polymers are touched and used by nearly everyone in one way or another practically every day. The rate of consumption of polypropylene has maintained the heady pace it set from the start and has set the record as the fastest growing polymer the world has yet seen (or appears likely to see). Starting decades behind the others, both polypropylene and linear polyethylene soon joined the select ranks of the 'billion-pound' plastics in the U.S. (shared only with polyvinylchloride, polystyrene, and their predecessor, high-pressure polyethylene).

The 'stereorubbers' and their progeny (ethylene-propylene co-polymers and tri-polymers: block co-polymers) have never displaced the old workhorse of the rubber industry, SBR (butadiene-styrene co-polymer), but they exceed in volume of production the combined total of all other synthetic rubbers.

Impact on science

The scientific impact has been of comparable magnitude. 'Ziegler/Natta catalysts' have been scrutinised by spectroscopists, rheologists, analytical chemists, physical chemists, inorganic chemists, polymer chemists, kineticists and semanticists to explain their unique activity, and have been tested in countless variations against every chemical molecule that could conceivably polymerise, plus many that could not. And the well is not yet dry; after two decades the exact structure and mechanism of action of these catalysts is still a subject for active research and lively debate*. We see here another example of the classical sequence: semi-empirical discovery, practical utilisation, theoretical explanation, the latter leading hopefully to new discoveries and a new cycle.

Economic impact—competition between Ziegler and Phillips processes

On the industrial/commercial wavefront, one crest was reached rather early, in the case of linear polyethylene. In a sense, it amounted to a victory of engineering over chemistry. It had seemed obvious to chemists and most others that the Ziegler process, operating at low temperatures and virtually no pressure, would have great economic advantages over the high-pressure polyethylene process with its special and expensive equipment and difficult reaction control. It also seemed simpler than the Phillips process, which operated at 500 pounds pressure in hot

---

*For example, Calvin Schildknecht maintains that the Z/N catalysts are of the cationic type, rather than anionic, as Natta claimed. Others feel that this is essentially a semantic, rather than basic, distinction. Reaction mechanisms and catalyst structures continue to be prime subjects for papers and discussion at meetings of the world's chemical societies.

solvent. But it turned out that the apparently formidable process costs were more subject to improvement than the cost of the chemicals used in Ziegler's process.

The high-pressure polyethylene producers continually reduced their costs through engineering advances, increased operating experience and ever-increasing scale. Their negligible cost for catalyst (initiator–oxygen or peroxides) then gave them the decisive advantage over Ziegler's process, for which the operating conditions were simple and inexpensive (once adequate catalyst handling procedures had been mastered) but for which the chemical (catalyst) costs were and remained substantial. Of course, the products of these two processes were so different that they did not compete directly in most markets.

The Phillips process, which makes a product very similar to Ziegler polyethylene, required only moderate pressure but had a difficult work-up step to recover the polymer from hot solution. Nevertheless, engineering improvements again dominated, and its low catalyst cost gave the Phillips process the advantage over Ziegler. Later on, still further process simplifications (the 'particle-form' catalyst) made possible even lower costs for certain grades of Phillips polyethylene. As a result, a number of Ziegler plants have been closed or converted to other products, and the Phillips process is now the leading source, worldwide, of linear polyethylene.

Because of the relatively short productive life of most Ziegler plants and the inordinate amount of product and process development required to launch the product, it is likely that most producers ended up with an unsatisfactory return, or even a net loss, when the interest value of money is considered. It is also likely that none made as much net profit on Ziegler chemistry as Karl Ziegler. It is unusual, but not necessarily deplorable, for the inventor to be the chief beneficiary of the fruits of his invention.

Montecatini's metamorphoses

In additon to licensing and joint ventures, Montecatini put up polypropylene plants of their own, but these were modest in scale and unimpressive in regard to product quality. Giustiniani continued to pursue an aggressive policy: cutting prices to capture markets, and litigation (or its threat) to enforce his (i.e. Natta's) patents and pressure other producers into taking a license. Not being familiar with U.S. business, marketing and licensing practices, he was somewhat at a disadvantage in executing his chosen 'mixed strategy' of entering polypropylene production in that country while at the same time granting numerous licences to other American firms.

Giustiniani was infuriated by the U.S. patent laws that deny foreign applicants the right to 'swear back' of their filing dates—a privilege that several U.S. companies exercised in attempting to invalidate Natta's patents in the U.S. He struck back with whatever means he could. Testimony has been presented to the effect that lawyers representing Montecatini induced an informant in the U.S. Patent Office to provide them covertly with confidential information regarding applications and statements filed by other polyolefin producers. Their deliberately obfuscating tactics in the interference proceedings have already been noted (Ch. 9).

Growing giants have giant growing pains, and Montecatini's patent battles, however wide-ranging, were not their main difficulty, nor polypropylene their main source of income in this period. Giustiniani's aggressiveness, abetted by Orsoni and, doubtless, others who shared or pretended to share his views, had pushed Montecatini into a string of expansions, acquisitions and new ventures of all sorts that stretched the company's financial and management capabilities to breaking point. It soon became clear that an infusion of outside capital and talent was essential to keep the giant alive.

Two abortive rescue attempts were made, the first being the formation of a joint chemical company with the Royal Dutch Shell Company. Within a year, it became evident that the two companies had such different approaches to both management and engineering that they were not compatible partners. It also became evident that this moderate venture could not generate cash fast enough for Montecatini's immoderate needs. The Italian company therefore supplemented the Shell joint venture by a merger with the Italian Edison Company, which had moved from the electrical industry into chemicals. This move caught Shell by surprise and created a *ménage à trois* which the directors found unacceptable. The Montecatini–Shell agreement was terminated by mutual agreement.

Formation of the new hybrid, Montedison, temporarily saved Montecatini, but not Giustiniani. The bankers had finally drawn the line at underwriting any more new ventures or bailing out old ones as long as he was head of the company. So Giustiniani disappeared from the scene, but Montecatini/Montedison's troubles did not. Losses continued and even increased under the supposedly tough-minded new boss,

Giorgio Valerio. But what sealed that president's doom was a telephone call from Eugenio Cefis, head of 'ENI', the state-owned petroleum industry. Cefis announced that he had secretly bought up enough Montedison shares to give ENI a controlling interest. A series of counter-moves by minority stockholders failed, as did a series of presidential successors (one lasted six months, the next, four). Inevitably, Cefis ended up as president of Montedison.

But the money haemorrhage was not staunched; Montedison lost a record $372 million in 1971 (heaviest of any of *Fortune* magazine's list of 300 largest industrials outside the U.S.) and still more in 1972. Heroic measures, such as pruning his accumulation of 1000 subsidiaries and 2300 products, adding certain other ventures, making deals with Russia and negotiating with the Italian government* enabled Cefis to bring Montedison close to the breakeven point. The price was semi-nationalisation: sharing control of the enterprise with the government.

However, even though these drastic actions and reorganisations probably, as Cefis claims, saved the company from bankruptcy, they did not eliminate its basic problems or keep it from sinking back into the red. Earnings again declined, and in 1976 Montedison lost $197 million, despite a large increase in sales. In 1977 Cefis resigned for the second time, this time unconditionally. He predicted that unless private shareholders subscribe huge additional amounts of capital the inevitable end will be complete state ownership of the company†.

After all these changes, scarcely anything recognisable remains of the organisations and people who brought polypropylene into the commercial world. But the plants, patents and problems inherited from the old company remain the responsibility of the new. So apparently, also remains the tradition of aggressive pursuit of goals set by the top man. Cefis, on taking over Italy's biggest—and sickest—corporation, acknowledged, in a Milanese metaphor: 'We suffer from too much individuality in this country. Everybody wants to be a *prima donna.* They all want to sing soprano, and nobody wants to sing contralto.'

---

*His negotiating techniques included resigning his chairmanship and withdrawing his resignation only when the government agreed to make the changes he demanded in policies affecting his company.

†In 1978 it was rumoured that a transfusion of capital lifeblood might be forthcoming from an unexpected source—a Middle Eastern banking consortium. Implications regarding resultant influences or control were not spelled out.

The litigations launched under Giustiniani's reign have also continued over the years under successive managements, with the end still not in sight. Fighting by their own rules but under another country's laws, Montedison have won most of the legal battles, but not yet the war. ('You Americans never know when to give up', one interested party complained.) To attempt a detailed analysis of the legal issues would exceed the scope of this narrative, the abilities of the author and the bounds of prudence. But the record left in their wake illuminates some of the events and the character of some of the principals.

Mission to Mülheim

Of particular interest among these records are the lengthy depositions taken in Mülheim and Milan to 'make a record' for reference in the U.S. patent infringement suits brought by Montecatini. Eight different corporate entities were involved in these actions, some of which have been consolidated, some not. Accordingly, it took quite a bevy of legal talent to represent all parties. A dozen lawyers, representing eight different legal firms, together with a court-appointed 'Master', arrived in Germany and settled in at the Duisberger-Hof, near Mülheim, on 9 December 1969, for the sole purpose of taking depositions from Karl Ziegler and his co-workers, none of whom were directly involved in the patent suits. Ziegler nevertheless had both von Kreisler and an attorney from his New York legal representatives present, as well as Heinz Martin, who was destined to be his deputy for licensing.

In the ensuing proceedings, which filled several weeks and over 1700 pages of transcript, both the power and the foibles of the legal machinery and profession were fully displayed. The sparring, the posing and the endless bickering by the opposing lawyers were occasionally clever, sometimes heated and rude, often petty and tedious, and outlandishly time-consuming.

'That the proceedings were tedious and lengthy is not surprising in view of the staggering documentation that had to be dealt with (a figure of 75,000 documents was mentioned at one point). To this load the transcript of the deposition, with its multitude of appended "exhibits", was to make a formidable addition. The fact that Montecatini (Montedison) was in the process of switching legal representatives in the U.S. meant that both law firms had to have representatives present, a fact that further complicated an already involved situation.

These preliminary skirmishes, like the court trials for which they set the stage, were battles of words, and the parties saw to it that there was no shortage of ammunition. Some of it was highly charged, if not of uniformly high calibre. The involved and sometimes heated exchanges led one of the participants to characterise the entire proceedings as "a three-ring circus".'

It must have been galling to Ziegler to be subjected to such grilling over an issue to which he was no longer a party but in which he might have had the principal interest had things gone as they should have some years earlier. Nevertheless, through all of the time-wasting sideshow, through endless hours of testimony reworking ground repeatedly ploughed on previous occasions, through repetitive questioning that was frequently impertinent or even insulting, Ziegler maintained a patient dignity. But that did not mean that they could push him around. One brief exchange is revealing:

'Which gentleman expressed this point of view?'

'Probably Mr Orsoni'

'You don't know who?'

'Probably Mr Orsoni.'

'Are you sure it was Mr Orsoni?'

'No, I am not sure; otherwise I would not say "probably".'

Had the Master been keeping score, that round would surely have gone to the chemist.

Despite his self-imposed restraint, Ziegler's testimony occasionally revealed flashes of his disillusionment and outrage over the actions that shattered his working relationships with Natta and Montecatini and had cost him the additonal credit and reward he felt he would otherwise have garnered. When the proceedings at last ground to a halt, the diplomatic Master, who had previously acknowledged the 'awesome powers of advocacy' of both attorneys for Montedison, spoke in German to Ziegler, thanking him for his voluntary assistance and for his 'attendance, his patience and his good humour throughout.' Ziegler made a polite response, but confined his expression of appreciation to the 'thoroughness' of the Master.

Mission to Milan

Three months later, the same cast, with minor substitutions, staged another circus in Milan, this time with Montedison in the dual role of plaintiff and host. A number of Natta's co-workers (Pino, Chini, Crespi, Magri, Mazzanti) testified regarding the Ziegler/Natta joint program and the discovery of polypropylene. Their testimony revealed, not surprisingly, that some disaffections had developed regarding the sharing of glory and compensation for that discovery. The lawyers tried to get something of this on the record by asking Pino about the rumour that Mazzanti's medal was larger than his. But Pino refused the bait; with stiff dignity, he would say of the other medal only that he had never seen it and of his own that it was 'of a certain size'

More significant were the questions some of Natta's people had raised regarding the naming of inventors on the patents and the fact that Pino had stated strongly to Natta his conviction that while it was nice to receive a medal as a token of his participation in Natta's scientific triumph, there was an entirely separate question of proper compensation for his contribution to Montecatini's commercial success. In short, he expected a bonus.

Natta had evidently referred the matter to Montedison, and Giustiniani, typically, had given Pino some reassuring promises but no money. However, in 1969, Pino was asked by Montedison to testify in the U.S. in connection with a patent interference case. And in November 1969, fifteen years after he had made his original request, Pino received from Montedison a letter acknowledging his right to a bonus and committing the company to a substantial payment in the near future. Left unsaid at the deposition that brought out these facts, but hanging heavy in the air, was the inference that these two events might have been related in more than time.

The proceedings dragged on for weeks, although it should not be inferred that the participants were dilatory. They sometimes started late, but they often ended late, and although they frequently observed the Italian tradition of the three-hour lunch break, they sometimes worked straight through, and also spent many hours between sessions in consultations in their hotel rooms.

The big question was whether or not Natta would appear and give direct testimony. His doctors and family objected, and the Montedison lawyers doubtless hoped they would prevail, but the defence lawyers persisted. Natta himself settled the issue by insisting on testifying. That action, his record and his demeanour induced all parties to treat him with respect and deference, in contrast to the near-insolence that had occasionally marked the questioning of Ziegler. Another major factor in the deference afforded Natta was his extreme debility, which required that the questioning sessions be very brief.

In his testimony, Natta verified the record of events and stated with quiet candour the reason for concealing his discovery of polypropylene from Ziegler.

'I did not tell him', he said, 'because I had to take patents first.'

No-one pressed the question of why he felt he had to act as he did. He was asked whether there was any prior art regarding 'stereoregular' polypropylene, and said:

'I don't think so.'

But when reminded of Schildknecht's polyvinyl ethers, he agreed that they were crystalline and stereoregular.

Natta was excused without cross-examination, and the Master, having expressed the thanks of the court to 'this honoured and distinguished man, who is obviously revered by his colleagues who know him well', closed the proceedings with the suggestion that they would appropriately be referred to as the 'Mission to Milan'.

Impact on the inventors

Both Ziegler and Natta were showered with honours, memberships and medals, and in due course, almost a decade after the fact, the discoverers of stereoregular polymerisations won the crowning accolade of the Nobel Prize. That recognition might have come sooner but for the dilemma that faced the Committee: should the prize go to Karl Ziegler, who was unquestionably the original discoverer but who made (or patented) only polyethylene, a variant on a known polymer? Or should it go to Giulio Natta, who took over Ziegler's information and applied it to making an entirely new polymer of propylene—the first one to be proven stereoregular?

The resolution was aided by a suggestion from another distinguished Nobel laureate (Sir Robert Robinson) who knew both candidates well. Like a scientific Solomon, he recommended dividing the prize between them, since their contributions could be considered equally important and distinct from one another, even though intimately related. And so it was done. In 1963, Ziegler and Natta appeared side-by-side on the platform at Stockholm to share the Nobel Prize in Chemistry.

From the language of the Nobel citations, it is possible, with a little effort, to recognise the distinction. Karl Ziegler's prize was for originating the remarkable catalysts and chemistry that bear his name; Natta's was not so much for the discovery of polypropylene *per se* as the brilliant elucidation of its structure and the elegant proofs of the existence of stereoregular polymers and helical chain structures in their crystals.

These awards naturally engendered a flood of congratulatory messages and complimentary articles, especially in the local presses. In the best universal journalistic tradition, the newspaper photographs in both home countries depicted their scientific heros in the company of attractive women: Natta receiving a congratulatory kiss from his sister and Ziegler dining with a princess and dancing with his grand-daughter in Stockholm.

The home-town accolades also followed local traditions. Mülheim organised a 'folk festival' that featured a beer wagon for the adults, sweets for the children (both furnished by the conferee), and a torchlight parade by the Institute staff, who in this fashion escorted Prof. and Mrs Ziegler back to their home. Later a second torchlight parade was held by the students at Aachen University.

In another colourful local celebration, Ziegler was ceremoniously conferred 'honorary citizenship' of Mülheim. In his address on that occasion, the Oberbürgmeister commented perceptively that Karl Ziegler would put the name of Mülheim on the map for the whole world (as a matter of fact, there are three Mülheims in Germany, but as far as scientists are concerned, Mülheim a.d. Ruhr is 'The' Mülheim).

Prof. Natta had received official word of the Nobel award while 'in retreat' at his father's home in Sanremo, which is close to the 'Nobel Villa' where the founder of the prize died. In an appropriate gesture, Natta immediately paid it a respectful visit. But the reporters quickly descended on him for interviews, even while he was still in San Remo. After all, he was the very first Italian chemist ever to win a Nobel Prize.

The writers turned out pages of purple prose about the 'Magician of Plastic Materials', the 'Faust of Matter' and even the 'Man Who Played God' ('*homme qui volait être Dieu*'). More restraint was displayed at the Polytechnic Institute. When Natta reappeared there, his colleagues had formed a reception group around his office door and greeted him with warm applause when he entered. Later, Natta gave a private party at his home for about forty people, including the Swedish Ambassador. It happened that this period also coincided with the Centennial of the Milan Polytechnic Institute, so a special double celebration was organised that took the form of (what else, in Milan?) an opera party at La Scala.

In recognition of the contributions of his key colleagues, Natta presented them with specially designed medals, which showed on one side the helical polypropylene molecule and on the other the seal in the Institute (which includes a detail from a sculpture by Leonardo Da Vinci).

Natta himself was presented by Montecatini with a coat made of polypropylene fibre, which he wore later when giving lectures on the subject.

It is also worth noting Natta's recognition of the contribution of Calvin Schildknecht in his early work on polyvinyl ethers (Ch. 2). Natta had referred to this work in his first publication on stereoregular polymers, and had corresponded with Schildknecht in 1948. In 1951, when in Europe on a lecture tour, Calvin and his wife were invited by Natta to visit Milan. Unable to accept, they had the invitation transferred to a colleague, J. M. Lambert (who had studied at the University of Vienna under a fellow-Austrian, Prof. Herman Mark) and his wife, who made the visit and were graciously received by Natta*. The invitation to the Schildknechts was repeated later, and he lectured at the Milan Institute in 1956 and again in 1973, when he presented a review of the history of isotactic polymerisation. By the latter date, however, Natta had been confined to his home for some time by failing health.

Research prospers—but the research captains depart

At Mülheim, the Max Planck Institute, with augmented prestige and funds, continued and greatly expanded its research program and facilities. Beside the ivy-covered, academic-looking original building arose a modern, multistorey, glass-and-concrete edifice that dwarfs the older structure and houses a much larger staff. Inevitably, the research is now more highly organised. While the scientific lightning has not struck again in the same place with equal force, the Institute's productivity remains high, attested by a continued series of developments of interest to the scientific and industrial communities.

In 1969, Karl Ziegler retired as director of the Institute and was succeeded by Günther Wilke. He no longer concerned himself with the day-to-day research activities, but continued to be the formal head

---

*It is thus obvious that Natta had been aware of Schildknecht's work all along, but apparently he had not immediately connected it with the unexpected crystallinity of polypropylene until reminded by reading Flory's book (Ch. 9).

of the Institute's licensing corporation (reflecting the long evolution of Prof. Ziegler, the organic chemist, into Dr Ziegler, the businessman). Even in licensing, however, daily affairs were handled by Heinz Martin.

Ziegler could now freely indulge his hobby of eclipse hunting, and on one occasion he even chartered an airplane so that he and his party could rise above the untimely overcast and escape being cheated out of their viewing. At home in Mülheim, he could enjoy making astronomical observations in the miniature observatory that had been constructed for that purpose on the top floor of the new laboratory.

But he had not relinquished the reins on major decisions, and when an important licensing question came to a head while Ziegler was on a world cruise, Martin had to cable him for guidance. Ziegler decided a first-hand discussion was called for, and told Martin to come out to wherever he could first intercept the Ziegler itinerary. That turned out to be Singapore, so Martin made an around-the-world trip to get his decision.

On the occasion of Ziegler's seventieth birthday, the Institute staff affectionately issued a commemorative booklet of cartoons, anecdotes and quotations, entitled simply 'Z'. In it are accounts of amusing minor incidents in Ziegler's career and numerous quotations that illustrate his own sense of humour, an attribute better appreciated, because better known, at home than abroad. This was also the occasion at which he endowed the Institute with the $10 million Ziegler Fund derived from his own patent royalties.

Ziegler made his final world cruise in 1972, taking his grandson along. His health declined soon after that trip, and he died on 11 August 1973, twenty years after the discovery that made him world famous, and just before his seventy-fifth birthday. The Max Planck Institute is now headed by Günther Wilke as research director and Heinz Martin as manager of the patent and licensing function.

The Milan Polytechnic Institute also savoured the fruits of success, and Natta continued the active personal direction of the research program as long as he was able. He was hampered by the onset of Parkinson's disease, which advanced so seriously that he had a series of brain operations in an attempt to arrest its progress. Even at the time of his Nobel Prize award (1963), he was seriously handicapped. One who was at the Institute at that time recalls Natta's coming to the laboratory every day, shuffling with painful care and slowness across the floor to his office, accepting only the minimum indispensable assistance.

Natta's voice, always low, became virtually inaudible. Yet, with great determination, he continued to work until it became physically impossible. He wanted to have still another operation, but was persuaded by doctors and family that it would be too dangerous. His wife having died some years earlier, he retired to live with his sister in Milan*.

Natta's office at the Institute is now occupied by one of his early colleagues, Prof. I. Pasquon. The office preserves numerous mementos of Natta's achievements, including a large painting that shows a statue of Allesandro Volta, with Natta climbing a huge spiral model of isotactic polypropylene in order to reach and shake hands with that great eighteenth-century Italian fellow-scientist.

We are left with the hope that Giulio Natta retained the satisfaction he voiced many years ago in answer to a question by a reporter.

'What satisfies me most? Having founded a school.'

Final respects: Karl Ziegler (1898–1973)

Karl Ziegler's death was noted around the world in obituaries that reviewed the career and enumerated the many honours of Germany's most famous scientist of two decades. His colleagues and students gathered at the Institute at Mülheim a week after his death for a memorial service. The first public meeting officially honouring his memory was held by the American Chemical Society, which had acted quickly to dedicate a symposium on coordination polymerisation, already scheduled as part of the April 1974 national meeting, to the memory of Prof. Ziegler. A few weeks after her husband's death, Frau Ziegler received a letter extending the Society's condolences and inviting her to attend the meeting to be held in his honour in Los Angeles, California.

She did not feel up to making the trip that soon and sent her regrets; however, Dr Wilke, now director of the Max Planck Institute, did attend and presented a review of Ziegler's life and scientific contributions. Piero Pino, now a professor at Zurich, also came to give a paper describing recent work on stereospecific polymerisation, as did others from Natta's 'School' and several scientists from American research laboratories where active pursuit continues of the many paths branching from Ziegler chemistry.

Karl Ziegler would have been gratified at this evidence of how far his work has been carried forward, yet without any sign of exhaustion of the subject, and he would have taken satisfaction at the fulfilment

*Note added in proof: Prof. Natta died in May, 1979.

of his hope that many younger scientists would enter and 'romp around', as he once put it, in the fields he opened. He would also have had a private smile for the outward amity with which the representatives of the Max Planck Institute and the Milan Polytechnic Institute participated in the Ziegler memorial program and in the animated panel discussion that closed it, free of any echoes of past bitterness.

In October 1974, the Max Planck Institute held a symposium honouring Ziegler's memory. Günther Wilke again gave a complete review of Ziegler's career and scientific work, and a series of papers was presented describing the continuing progress in the many lines of research which he had initiated.

12 Shadows of the Future

> What had exclusively been in the power of Nature, namely to join the monomeric units in a giant molecule with a predetermined steric order and not at random, is now possible also for man. He has even created types of macromolecular structure that do not exist in nature.—Giulio Natta

As a result of the trail-and-error process by which new products elbow their way into some fields, are weeded out of other fields and capture unoccupied territory, the older plastics and other materials eventually made room for the newcomers whose births we have attended. New technologies (e.g. fibre making by slitting and stretching film), new products (e.g. biaxially oriented film) and new uses (e.g. outdoor-indoor carpeting) aided the young family of 'stereopolymers' to achieve that unprecedented growth rate that seems destined to make them the dominant class on the plastics scene*.

The mark of maturity and the price of pre-eminence for either polymers or people is the growing necessity to be concerned over threats of displacement by the next generation of upstarts. Unlike people, however, older polymers do not often fade entirely away, but simply surrender some of their ground and their future prospects to the new contenders. So the question is not, 'What will displace stereoregular polymers?', but, 'What will come along that will demand a share of their place in the sun?'

---

*Polyolefins as a class (including high-pressure polyethylene, but not polydiolefins) have been forecast by Stanford Research Institute to reach a world volume of 43 million tonnes by 1980, compared to 20 million tonnes for vinyl polymers and 5 million tonnes for polystyrene.

What seems remarkable, in fact, is that after the explosive burst of discoveries in the mid-fifties, when significant inventions followed one another only weeks and months apart, nothing has appeared in the ensuing twenty years that shows any promise (or threat) of becoming another large-volume, general-purpose plastic or synthetic rubber. The same is true for fibres, where there have been no really major new entrants since acrylics and polyesters (except for polypropylene, which is not quite yet a 'major' fibre).

With a few debatable exceptions, the new polymers that have been introduced are 'engineering thermoplastics', speciality rubbers, or other high-performance materials that serve special uses but are not versatile enough or low enough in cost to compete across the board. The potential exceptions, cost-wise, are themselves second-generation stereoregular polymers or progeny of that technology. High-performance polyolefins have been made from higher olefins (1-butene, 4-methyl-1-pentene). Polybutene and stereospecific polystyrene (isotactic or syndiotactic) are intriguing because of attractive properties and potential low cost, but they have serious drawbacks, particularly with respect to phase changes, that have so far not been overcome.

There is a strong temptation to say that there is simply no 'room' for another general-purpose polymer, since all needs are satisfied by the large and versatile family of materials already available, and that zooming raw-material and construction costs would put any newcomer at a hopeless disadvantage. We must bear in mind, however, that all this was said before—in fact, just before the discoveries that made stereoregular polymers and this story possible. Scientific discoveries and inventions of equal or greater moment may be made at any time. What is far less likely is that they will encounter the precise combination of circumstances that will turn discovery into fruitful innovation of comparable magnitude to that which we have described.

Accepting the fact that such events are unpredictable, we may note a number of current factors that worsen the odds. To the 'old timers' among polymer chemists it seems obvious that the 'golden age' of polymer research is past, as it is even for organic chemistry generally. True, there are more people working on polymers and in better facilities than ever before, but that fact constitutes primarily further support of Prof. Parkinson's (of 'Parkinson's Law') hypothesis regarding the proliferation of work long after the initial creative burst. Pulses throb more strongly today over the challenges in molecular biology, 'environmental science' and the social sciences.

On the industrial front, the play has moved dramatically away from new products and new markets and has become a defensive game with the primary and worthy goals of protecting the public's environment and the company's rear (raw-material shortages, legal compliances, etc.). Aggressive moves are largely limited to trying to capture a share of an established market by offering an established product at a competitive price.

Research, obediently, has also become defensive and process-oriented, concerned with new catalysts that will perform known functions better and with process engineering that will reduce the cost of a known operation. With the long-overdue recognition of the necessity of conserving energy, natural resources and the ecosphere, and with the economy increasingly emphasising services more than goods, the hurdles now placed before any major new product venture in polymers have been materially increased in number and raised in height.

What we can clearly see (and therefore foresee continuing) is a trend to meet needs and opportunities in the materials field by modifications and combinations of existing materials, rather than by introducing entirely new materials. 'Materials Science', itself a relatively new discipline, has become enamoured of 'composite materials' whose permutations and combinations are mathematically limitless.

Even the new polymers of current greatest interest—graft polymers (made by chemically bonding groups of different composition to the back bone of a preformed polymer molecule) and block co-polymers (. . . –AAAAAAAA–BBBBBBB–AAAAAAA– . . .) are combinations or modifications. And one of the boldest prognosticators of them all, Herman Mark, has predicted that the most promising area for advances in polymers is in combining the technologies of thermoplastic (meltable) and thermoset (irreversibly heat set) polymers in a single product.

It appears safe to predict also that techniques will be found for achieving stereospecific polymerisation in aqueous (emulsion?) systems; indeed, considerable research has already been devoted to that end and with some degree of success. Equally safe is our anticipation of great progress in understanding the crystal habits of microcrystalline polymers and their strong effects on properties, to be followed by greatly improved control of such crystal structures to achieve different and desirable products. Orientation of crystalline polymers by forced deformations below the melting point has produced striking effects and useful products when carried out in one dimension (fibres) and two dimensions (biaxially oriented film); is it not possible that means will be found to orient in three dimensions and with useful results?

Developments of the sort just named are predictable because one can visualise a theoretical approach, an extension of known principles or technology, or a broader application of known products; in short, 'a way to get there from here'. But we would be unwarranted in assuming the obverse: that things for which we can visualise no means of attainment are necessarily unattainable. In Perry's *'Chemical Business Handbook'**, the chapter entitled 'Research' contains the following passage.

'A research program should have an attainable purpose, that is, means should be available by which a solution may be found and the solution should be such that action can be based upon it. The problem of determining the topography of the far side of the moon, for instance, is not suitable for research since there is no known means for attacking the problem.'

At the time those words were written, Russian scientists were busily engaged in research aimed at that precise objective, with subsequent results that startled and impressed the world. Despite any contextual qualifications the authors of the quotation might invoke, they have given us an excellent example of the dangers of being too cautious in predicting what research can and cannot accomplish in the future. With this lesson in mind, the only prediction that will be ventured in these pages is the safe one that there will be important developments in polymer science and technology from entirely unexpected sources, and from research aimed at other goals.

There is available a less inhibited and more eloquent prediction of the future impact of the advent of stereospecific polymerisation, from a more impressive source—the man who originated it. Here is Giulio Natta's own conclusion†.

'A revolution will be marked . . . by the development of processes that lead to the formation of macromolecules having a predetermined

---

*Perry, *Chemical Business Handbook,* McGraw-Hill, New York, p.3 of Ch. 4 (1954).

†Natta, G., 'Present Situation and Prospects of the Italian Chemical Industry—High Polymer Developments', *Can. J. Chem. Engng,* **114** (1948).

ture. They will render some branches of industry independent of agriculture (in the fields of textiles, of paper and of building materials) and increase the areas of land used for the production of food. In a few decades this will be strict necessity dictated by the impressive growth of world population and by the increase of the standard of living which is the aim of all nations. Our peeping into outer space cannot distract us from our earthly problems.'

13 Interpretations

It is easy for imagination to be eloquent but hard for it to be fair. No moralist should write history.—Will Durant

Without moralising, we may profitably study history to select those parts we should like to see repeated. Surely the phenomenal burst of fruitful innovation we have just recounted represents the sort of thing that scientists, research directors and business managers would like to think could be induced to occur again in similar or different fields. Judging by the recent literature on the subject, a great deal of thought and effort is being expended to find and nurture the seeds of creativity and innovation.

In what is perhaps the most ambitious of such attempts to date*, U.S. President Nixon in 1972 charged the National Science Foundation and the National Bureau of Standards with making an in-depth investigation and report on how the government could 'improve the climate for technological innovation'. This undertaking was slow to get off the ground, suffering from an apparent inability to come up with an innovative approach (!). But in 1973, Battelle Research Institute issued, under National Science Foundation sponsorship, a report that employed the case-history approach to derive a number of fairly orthodox conclusions about the desirability of encouraging pursuit of unorthodox ideas, the advantages of interdisciplinary teams and the key role of the 'technical entrepreneur'. Of particular interest to us is the fact that the time span from conception to realisation in their examples averaged nineteen years. By contrast, we are forced to conclude that the startling speed

*The latest study was launched by the Carter administration, this time at Cabinet level, but results are not yet apparent.

with which fruition followed first findings in the cases we have been examining (about five years) must reflect an exceptionally favourable set of circumstances.

Case histories, like analogies, can be carried too far. But, since not even the N.S.F. has been able to conduct any properly controlled experiments on innovation, examination of past events affords the best present means of applying the test of real experience to hypothesis and conclusions. In view of the outstanding success and unusually rapid fruition of the discoveries of stereoregular polymers, it should be instructive to look for patterns and key factors among the numerous events that make up their history.

A polymerisation reaction can be carried out successfully only when the right ingredients are present under the right conditions. It is then characterised by three distinct phases: (1) initiation, (2) propagation, and (3) termination. Similarly, 'fruitful innovation' will occur, in polymers or any other field, when conditions are favourable and it, too, involves (1) initiation, i.e. a discovery or invention, (2) propagation, i.e. the additional research, development, engineering, manufacturing and marketing that transform a discovery into goods and services useful to mankind, and (3) termination or, more usually, the inevitable levelling off when needs are filled and the decline that occurs when newer products or processes are found to serve those needs better.

Our concern is with factors that can favour the first two of these phases and perhaps postpone the third. We are also interested in recognising which of these factors are subject to deliberate control or at least to influence, and which are not.

It will be illuminating for our purpose to review in summary form the discoveries that have figured in our story and to note which represented expected and which unexpected results. We shall also note which were pursued to a significant outcome, and therefore can be cited as examples of fruitful innovation, and which remained relatively (or entirely) unnoticed. Finally, we shall look for general characteristics, such as the conditions, timing and environments, that are common to those of the discoveries that were carried through to practical and commercial successes.

Let us look, therefore, at an abbreviated chronology:

1900–Von Pechmann makes polymethylene from diazomethane.

An unexpected result, obtained long before the era of high polymers; hence no major impact.

1929–Marvel observes polymerisation of ethylene by lithium alkyls.

An entirely unexpected discovery, it had little or no impact because the process was not efficient and because polymer interest and know-how, even in duPont, were still embryonic.

1933–Gibson and Fawcett discover high-pressure polyethylene. Entirely unexpected, this discovery was made and had tremendous impact because of several fortuitous events, alert observation and aggressive development.

1937–Hall and Nash observe formation of aluminium alkyls during polymerisation of ethylene with aluminium chloride. Potential significance of this unexpected result was not recognised; hence no follow-up and no impact.

1940–Ziegler describes polymerisation of butadiene with lithium alkyl. Little impact because conditions did not favour high molecular weight or high-*cis*-1,4- content and no analytical means were available to observe *cis* content.

1940s–Partially crystalline polyethylene made in U.S. Rubber Co. laboratories but not pursued because evaluation of product was pessimistic.

1942–Polymerisation of ethylene passed over cobalt catalyst in Shell Development Co. laboratories is unexpected nuisance. No follow-up; hence, no impact.

1943–Fischer polymerises ethylene with aluminium chloride plus titanium. A low polymer was expected, but not the solid, possibly partly crystalline product actually obtained. No significant impact, presumably due to lack of follow-up by BASF.

1947–Schildknecht observes unexpected formation of both crystalline and amorphous polymers from vinyl isobutyl ether and ascribes existence of two forms to stereoisomerism. Impact significant but moderate, due partly to limited commercial importance of product.

1953–Ziegler (Holzkamp) makes crystalline, linear polyethylene using transition-metal catalysts. This unexpected and partly fortuitous discovery, with subsequent related discoveries and development, had enormous scientific and commercial impact, due to aggressive follow-up and ideal timing.

1953–Foster and Wakefield succeed in efforts to make high-*cis* polybutadiene and polyisoprene, using lithium catalysts. Pursued aggressively internally but secretly, these discoveries made Firestone first with a high-*cis* polydiene rubber, but impact was diminished by delay in public announcement and decision not to make polyisoprene commercially.

1954–Natta (Chini) makes crystalline, isotactic polypropylene. Rubbery polymer expected; crystallinity unexpected. Brilliant follow-up research and aggressive development by Montecatini generated enormous scientific and commercial impact.

1954–Ziegler (Martin) makes solid polypropylene, following earlier inconclusive experiments. Later he learns that Natta has made the same discovery and has an earlier patent application date.

1954–Hogan and Banks make linear polyethylene while trying to make gasoline. Aggressive pursuit of this entirely unexpected finding led to Phillips process becoming dominant worldwide.

1954–Rehn (Hoechst) makes polypropylene; Hoechst defers development in deference to Ziegler.

1954–Zletz (Standard of Indiana) makes linear polyethylene while attempting an alkylation reaction. Relatively non-aggressive development by Standard resulted in later and less extensive commercial success than Phillips process.

1954–Vandenberg (Hercules) makes solid polypropylene using Ziegler catalyst. Hercules went on to become pioneer U.S. producer of polypropylene.

1954–Horne (Goodrich) makes 'synthetic natural rubber' (*cis*-1,4-polyisoprene) while trying to make an ethylene-isoprene co-polymer. Immediate recognition of the significance of this unexpected result and aggressive pursuit achieved major scientific impact and ultimately the best commercial position for B. F. Goodrich.

1955–Lippincott (Standard of New Jersey) makes polypropylene and isolates a solid fraction. Other chemists at Standard later made improvement in catalyst preparation techniques that were an important aid to successful commercial development.

1955–Porter succeeds in efforts to make high-*cis* polyisoprene and polybutadiene using lithium alkyl and cobalt catalysts. Shell Chemical developed both processes but gave polyisoprene first priority and became its first commercial producer.

1955-present–Innumerable additional catalyst, process and product improvements by countless other workers have developed stereoregular polymers into the world's largest and fastest-growing class of high polymers.

Several things stand out immediately in this recapitulation.

(1) All the discoveries that had major impact were made in large, well equipped industrial or industry-sponsored laboratories. A number of relevant 'prediscoveries' of potentially high impact were made in

university laboratories but went relatively unnoticed and hence undeveloped. This fact does not establish the superior creativity of industrial chemists. Rather, it is clear evidence that fruitful innovation occurs most often when the research, engineering, development, product application and market development capabilities that a major industrial organisation can muster are fully deployed to convert a scientific breakthrough into a new commercial process and product*.

(2) Most of the discoveries represented results altogether different from those being sought, and in all but a few instances the result was surprising and unexpected. In only two cases, in fact (Foster and Wakefield, Porter), did the inventors accomplish precisely what they set out to do. The unexpected result, moreover, usually proved to be of much greater importance than achieving the original target would have been.

(3) Fortuitous events and, particularly, fortuitous timing, played a major role in an impressive number of cases. This fact and the previous point could be (and have been) interpreted as an indictment of 'targeted' or 'budgeted' research. But I would submit that all good research is 'targeted' in the sense that the experimenter has *some* goal toward which he is working (none of those who figure in this history was an aimless putterer!), and that it was flexibility in modifying targets in the light of new findings that was an important key to success in the cases at hand. Alertness to note, curiosity to pursue, and intelligence to interpret the unexpected result made the difference between trivial and tremendous consequences.

Whether research targets are chosen by formal or informal processes, or even subconsciously, the critical choice is between targets that are 'scientific' or 'practical', small and easy or large and difficult, and those choices must be made on the basis of the kind of gamble that the scientist and his sponsors are prepared to take with his career and their resources.

(4) In every case we have examined, fruitful innovation occurred only when there was effective communication, both internal and external. We saw one striking negative example in which priority and initiative were lost because knowledge of a discovery was withheld from top

---

*Karl Ziegler's Institute did not have all these capabilities in house, but that lack was amply compensated by his capability for inducing others to supply them.

management (Firestone, Ch. 10); and undoubtedly the course of every other development was affected, favourably or otherwise, by the quality of communications, upwards, downwards and horizontally, within the organisation. We have also noted in several cases the beneficial effects of frequent and direct communication between the 'man at the bench' and the research director. (This does not fit commonly accepted views, or my own, as to how research should be managed, but the facts cannot be denied.)

But *external* communication has played an equally important role. The rapid dissemination of news of a discovery through public and private contacts brought about a tremendous proliferation of interest and activity among both research and 'commercial' people. The 'first to know' often won a ground-floor opportunity, even though it was not always seized.

Noteworthy in this respect is the key function that scientific meetings repeatedly performed in this vital communication activity. Those concerned with fostering creativity—and those managers who sometimes regard attendance by their staff at professional meetings as a luxury or a 'boondoggle'—should find it instructive to trace in the events we have chronicled the striking frequency with which an American Chemical Society meeting, a Gordon Conference or a European scientific conference has provided the source or the occasion for a vital stimulus. It came in different forms on different occasions: a significant paper heard, an idea tested informally on colleagues, an exchange of news or data, an encouraging word from a respected peer, a personal contact made or renewed—but the common result was that research momentum and enthusiasm were created that led ultimately to scientific and practical successes, i.e. to fruitful innovation.

(5) An unprecedented burst of discovery occurred in one remarkably brief period (1953-55). This was partly the result of the spreading and cross-fertilisation of new ideas in a newly opened field (Ziegler chemistry), but other discoveries (at Firestone, Phillips, Standard of Indiana, for example) were being made simultaneously and essentially independently. The other major favourable ingredient in the situation was undoubtedly timing. There is a tide in technology as in other affairs of men, and the successful individuals and organisations managed, by good design or good fortune, to take it at the flood.

As far as tools, technology and facilities are concerned, there is no reason why regular, and even stereoregular, polymers could not have been discovered as early as the late nineteenth century. The laboratory techniques, while demanding, were well within the capabilitities of turn-of-the-century laboratories, at least in Germany. The reagents used, although then highly exotic, could have been synthesised, and the equipment requirements were modest. The work did not have to await the development of nuclear magnetic resonance spectroscopy, or molecular orbital theory, or computers, or ultrahigh-pressure apparatus.

In fact, as events have amply demonstrated, it would have been perfectly possible, at least twenty years before the actual discovery, for an imaginative chemist to have predicted and then proceeded to prove (given some luck with his experiments) the existence of stereoregularity in polymers. But we must recognise that, had the discovery been made twenty years earlier, it could not have won the recognition and rapid development that in fact occurred, for neither science nor industry would have been able to recognise or appreciate the uniqueness and virtues of the product.

What made Ziegler's discovery epochal instead of incidental was timing. Thirty years of painstaking build-up of the scientific understanding of long-chain polymers and its associated technology had finally created an environment in which the distinctive fundamental characteristics of a new polymer could quickly be recognised and its practical merits quickly assessed and exploited.

Moreover, in the mid-fifties the technological tide was augmented by the strong wind of the business climate, which was sweeping all enterprises in the direction of expansion, new products and new ventures. A decade later the business winds had reversed: new ventures became much less popular, and retrenchment in research became the order of the day. Had Ziegler's and Natta's discoveries been delayed ten years, they would have been greeted with far less enthusiasm, perhaps none.

(6) Certain individual and institutional names reappear with impressive frequency in this history. We thus see one more illustration of the fact that a few key people (people being the vital force of institutions) often have a disproportionate effect on events far beyond their immediate sphere of activity. Besides Karl Ziegler, Giulio Natta and their respective institutes, we have repeatedly cited such names as Hermann Staudinger, Herman Mark, Sir Robert Robinson, the Max Planck Institute, The American Chemical Society, Johns Hopkins University, The University of Chicago, Brooklyn Polytechnic Institute, etc. Without their influences, the same discoveries might still have been made eventually, but certainly much later and with far less impact.

Fostering fruitful innovation

On the basis of this limited sampling, what can we say are the vital elements that must come together to make fruitful innovation possible? They may reasonably be grouped into four categories, the first of which deals with the setting for, and the last two, the consequences of, the central event—the initial discovery.

1. Man, mind, means, motivation

The skilled and knowledgeable experimentalist with a trained and enquiring mind is the first requisite for the kind of discoveries we are considering. To repeat an observation made earlier, the inventor is the indispensible initiator, but for his invention to become a case of 'fruitful innovation' requires a long chain of subsequent events and an equally indispensible succession of innovations by those responsible for each link in the chain.

2. Discovery

A significant experimental result must be observed, whether unexpected (usually) or predicted and confirmed (rarely). As Pierce* has said: 'In technology as in science, the individual act of creation is the essential ingredient of innovation. It may be an act of creative imagination . . . or an act of creative observation'.

3. Follow-up

The discoverer must have the alertness to recognise the novel result and the scientific curiosity and the will (also, the freedom!) to follow where it leads, even if away from the preconceived path. Deciding when a bypath should be pursued and when passed-by as a dead-end is probably the most critical judgement factor in research. Once undertaken, the follow-up research requires good experimental techniques and good judgement in planning and carrying-out experiments. We have seen instances where lapses in this regard proved very costly.

But beyond the laboratory phase, a lengthy sequence of additional efforts by many people is required if the initial discovery is to be converted into a major contributor to human needs. Typically, this involves

---

*Pierce, J. R., 'Innovations in Technology', *Scientific American*, Sept. 1958, pp 116–130.

research and engineering to develop a practical manufacturing process, extensive product evaluation, and the final test of usefulness to people—will people buy it?

4. Fortune

Only a fraction of all well conceived research programs result in a significant discovery, and of these only a fraction turn out to have major social and economic impact. An invention made before its time languishes; one made too late has already been passed by. Optimum timing involves both the right degree of technological and cultural 'ripeness' and the right economic environment—a business cycle on the upswing, an expansionist mood in industry, a recognised need or one that can be stimulated.

All these points have been made by others and have been discussed more extensively; the purpose of restating them here is to raise the final question: What can be done about them? Can they be controlled or strongly influenced in directions favourable to fruitful innovation?

Obviously, something can be done about the first set of factors: *men and means.* The scientist can train and discipline himself; organisations can (and do) devote great energies to seeking out and selecting the best brains and can provide them with the best facilities. But no-one can command *discovery.* The best that can be done is to make the chances as favourable as possible, be alert for the unexpected event and take full advantage of it, believing with Pasteur that 'Chance favours the prepared mind.' His observation is as pertinent in the twentieth century as in the nineteenth, and has been made a truism by being repeatedly confirmed, quoted, plagiarised and paraphrased. Shimkin* has recently put it more explicitly: 'The basis of scientific discovery is the prepared mind confronted by new or unexpected findings.'

The essential *follow-up* activities can be provided by a major industrial organisation; the most important and the most difficult to provide is good judgement as to which projects to back to the hilt and which to drop. Halfway efforts produce equivocal results; all-out efforts on an unsound project can produce disaster.

The good *communication*, without which there is unlikely to be any effective follow-up, can certainly be strongly encouraged; however, as

*Shimkin, *Science,* May 18, 1973, p 693.

many a previous study has concluded, it cannot be commanded. As with discovery itself, communication depends mostly on the character of individuals.

To quote Pierce again:

'Yet, while insight and invention are the products of individual minds, individuals are most effective when they work together in a field that is neither so narrow as to preclude adventure nor so broad as to scatter energies and prevent a fruitful interchange of ideas.'

In the same vein, we may add that managers of research will do well to see to it that neither research assignments nor management's reins are so broad and loose as to scatter energies nor so tight and narrow as to stifle individual initiative and creativity.

Finally, it goes without saying that the smile of *fortune* cannot be invoked, but only hoped for, in regard to either discovery or its timing. Forecasts of needs and of business cycles can be obtained from planners or astrologers, often with results only marginally different. Much has been said and written about 'Planned Creativity', and enthusiasts are even capable of such statements as: 'The historians of the future may well select the development of deliberate creativeness as the most important development in this century.'* However that may be, present historians must surely conclude that neither creativity nor the more extensive sequence I have labelled fruitful innovation have yet been successfully planned, and can at best only be encouraged. The fact is that no more than *half* of the critical factors enumerated above are subject to deliberate control.

We are left, then, with the conclusion that the most that individuals or organisations can do to favour fruitful innovation is to have the best possible minds at work, adequate facilities, freedom to pursue intriguing new leads, good communication opportunities and ample follow-up capability—but all this is still not enough. The additional requirement is uncontrollable but vital—'a little bit o' luck.'

---

*Nelles, in *'The Chemical Bulletin'* (American Chemical Society)

Epilogue

It seems fitting to end this story with a tribute to the men who created it. Most of the principals now represent an older generation of scientists and the work we have described has already passed into history. But its significance remains, for science and technology always stand on the shoulders of previous generations, and the half-forgotten innovations of the past have built the platforms upon which the future will be reared. The world will not be the same for these men having lived and worked in it, and it will have need of their breed as long as man is to survive. We do well to consider how society can continue to generate, support and encourage such innovators in the future, that we may thereby continue to have a future.

Index

**Note:* '(Q)' denotes a quotation and '(R)' a reference